Eurípedes Moreira de Melo
Hérika Patrícia Fragoso

The (wrong) paths of tourism

Eurípedes Moreira de Melo
Hérika Patrícia Fragoso

The (wrong) paths of tourism

Estrada-Parque Prefeito Divaldo Rinco - GO - 239

Imprint

Any brand names and product names mentioned in this book are subject to trademark, brand or patent protection and are trademarks or registered trademarks of their respective holders. The use of brand names, product names, common names, trade names, product descriptions etc. even without a particular marking in this work is in no way to be construed to mean that such names may be regarded as unrestricted in respect of trademark and brand protection legislation and could thus be used by anyone.

Cover image: www.ingimage.com

This book is a translation from the original published under ISBN 978-613-9-70240-4.

Publisher:
Sciencia Scripts
is a trademark of
Dodo Books Indian Ocean Ltd. and OmniScriptum S.R.L publishing group

120 High Road, East Finchley, London, N2 9ED, United Kingdom
Str. Armeneasca 28/1, office 1, Chisinau MD-2012, Republic of Moldova, Europe
Printed at: see last page
ISBN: 978-620-8-17470-5

ACKNOWLEDGEMENTS

I thank God for the physical and mental health he is able to give me. To my sisters for their support in this madness that is studying. I would like to thank my friend Gilvan Andrey for the relaxation time, and my friends and bosses at PROEX/IFG for the words of encouragement and affection with which they treated me throughout my internship. Special thanks to Professor Gisélia Carvalho for the great energy she transmits. I would also like to thank Professor Carlos Shiley, my supervisor, and Professors José Carlos and Clarinda, who agreed to sit on our panel. Finally, a huge thank you to Ivo Augusto for teaching me Buddhism, so that I could believe in everything a little more.

Eurípedes Melo

SUMMARY

This study focuses on the environmental aspects caused by the paving of the GO-239 (Estrada-parque Prefeito Divaldo Rinco), as well as an analysis of the aspects related to the signposting of this road, along the route between the district of São Jorge and the city of Alto Paraíso de Goiás. In it, we highlight the fact that the GO-239 is called a park road, which in theory should have wildlife corridors (tunnels for animals to pass through safely), speed reducers to 40 km/h at points where animals are seen crossing, vertical signage and other requirements. In order to find out whether these issues exist on the motorway, a study was carried out in the area. The 36-kilometre stretch of paved road linking Alto Paraíso to the São Jorge district was travelled several times to analyse the asphalt, the signage, the cycle path and the tourist attractions found along the route. The research, which uses the qualitative method, aims to consider the answers received by looking at their particularities and delving into the individuality of the interviewee's thoughts. To this end, residents, shopkeepers, tourist drivers and the administrator of the Chapada dos Veadeiros National Park were interviewed in the community of the São Jorge District. This road has great tourist potential given the scenic beauty that surrounds it. It is a park road that exists in law, but not in fact. We believe that if the administrators, environmentalists and even the community got together to draw up a management plan that favoured this stretch of road, highlighting its real value, it would become an added attraction for the region.

Keywords: Park road, Tourism, Asphalt and Chapada dos Veadeiros National Park.

SUMMARY

CHAPTER 1

INTRODUCTION

Natural beauty and esotericism bring tourists to São Jorge in search of the rest, peace and quiet that the place offers. The easier it is to reach the village, which is 36 kilometres from Alto Paraíso de Goiás, the greater the flow of people. If the number of people increases, the wear and tear on the place increases even more, causing the local community to suffer as a result of the increase in tourism.

Located in the north-east of the state of Goiás, the town of Alto Paraíso de Goiás is 230 kilometres from Brasília and 460 kilometres from Goiânia, and is internationally known for its beautiful waterfalls, as it is here that the main entrance to the Chapada dos Veadeiros National Park is located.

This district has approximately 500 residents, but that number rises to around 4,000 if tourists are added during high season. However, with the completion of the paving of the GO-239 in 2014, the number of tourists in the town has increased considerably. According to Dias (2005, p. 55), the reason for a tourist's visit to a particular place "is fundamental for identifying the visitor's consumption and spending behaviour".

This paper presents the results of an analysis of the GO-239 motorway, which was initially called the Sulivan Silvestre Park Road and is now called the Prefeito Divaldo Rinco Park Road.

With the completion of the asphalt paving and improved signage in 2015, since everything was very precarious before that, the 36 kilometres of this road, between São Jorge and Alto Paraíso de Goiás, were observed and photographed in order to find out why this road has been given the title of "Park Road", since it doesn't have a management plan, and how this could be proven through field research.

Given that there is very little literature on the subject, studies by various authors were collected from social networks, and various academic works were used. These include an extensive study on the subject by Soriano (2006). The authors Dourojeanni (2003) and Tricário, Oliveira, Rossini and Carvalho (2012) were also used.

In Chapter 3, we emphasise various concepts about what tourism is and how it has evolved over time. For the theoretical framework, we highlight three authors who have studied the subject, namely Dias (2003, 2005), Rejowski (1996) and Solha (2002). This chapter also analyses the tourism practised in Brazil between 1960, when studies on the subject began, and 2016.

Chapter 4 deals with protected environmental areas and their objectives, Conservation Units (CUs) and their categories, as well as the landscape in these areas. Here we draw on Dias (2003, 2005) and Hauff (2004).

The various definitions of what park roads are, which have this name because they are not ordinary roads, i.e. they are roads characterised by their unique aspects in archaeological, cultural and historical terms, as well as having other features such as natural and recreational areas and high scenic quality (NATIONAL SCENIC BYWAYS PROGRAM, 2009), is dealt with in Chapter 5.

This same chapter also covers the creation of Park Roads in Brazil and a brief presentation of the main roads that already have legal backing, including the Romantic Road in Germany, the MS-184 and MS-228 state highways, in the Pantanal, MS, the Serra da Cantareira Parkway in São Paulo, the Itacaré Parkway in Bahia, the RJ-163 and RJ-151 between Visconde de Mauá and Capelinha in Rio de Janeiro, and finally the Graciosa Parkway in Paraná.

Chapter 6 presents how the city of Alto Paraíso de Goiás and the District of São Jorge, which is the main gateway to the Chapada dos Veadeiros National Park (PNCV), came about. The authors used here were Behr (2000), Domiciano (2012, 2014), Silva (2003) Silveira (1997) and NSUC (2000).

Chapter 7 provides a survey of the paving of the GO-239, which is the subject of this work, focusing on the environmental aspects and signposting of the road, considering the definitions of what a Parkway is and why it is so called.

The field research carried out on the road, along the route between the village of São Jorge and Alto Paraíso de Goiás, with the aim of gathering the opinions of residents, tourist drivers, shopkeepers, local tourists and the PNCV administrator, is explained in Chapter 8.

Within the qualitative method, our research is also exploratory in nature. According to Gonçalves (2014), "this aims to discover, find, elucidate phenomena or explain those that were not accepted despite being evident".

The sample taken from residents, traders, tourist drivers and administrators working in the region was done by saturation, i.e. "that which denotes, during the analysis, that no new properties or dimensions emerge in the data, and that the analysis accounts for a large part of the possible variability" (STRAUSS, 2008).

CHAPTER 2

RESEARCH METHOD

The research carried out in this study is qualitative in nature, as it takes into account the opinion of the subject interviewed, using results obtained through interviews with questionnaires or scripts that aim to provide answers to characterise the place being researched.

According to Anjos (2004), the purpose of data collection is to gather and record all the data relating to the object of study, thus providing a clearer and more objective form of the subject, clarifying the theme of the study.

Unlike quantitative research, which seeks to boost the number of interviewees for a more numerical reading, qualitative research makes use of specific criteria cited by authors who live the phenomenon studied every day. Thus, in this work we prefer to use local knowledge to form a text that comes closer to the reality experienced by the residents of Vila de São Jorge, since it is in this region that our case study is focused.

a) RESEARCH DEVELOPMENT AND RESULTS

This research was carried out over two time periods. Firstly, a theoretical framework was drawn up, starting with the emergence of tourism up to the present day, going through the history from the 1960s to 2016. The second part of the study focuses on the emergence of Park Roads and the standards used to designate them as such.

In order to reinforce the designation of the GO-239 as a park road, between 11/02/2017 and 15/02/2017 we carried out field research on the stretch of asphalt linking Alto Paraíso de Goiás to São Jorge, a route totalling 36 kilometres.

To analyse the start of the paving, we continued for a few more kilometres, stopping at the main points along the road to analyse the asphalt, the cycle path, the signposts, the viewpoints and more.

We used a number of technical procedures to construct our work, including bibliographical research to support our work and documentary research taken from books and academic works written to support our text, as well as a survey and field research with interviews using pre-assembled scripts.

The trip to Vila de São Jorge lasted four days, with one day dedicated to studying and analysing the road and the others to interviews in the community.

The route was travelled round trip, with strategic stops at tourist attractions, roadside viewpoints (Maytrea and Buracão), wildlife corridors and speed bumps.

Photographic records were taken of all the signposts along this stretch, as well as the tourist attractions and the condition of the road.

The interviews, in the form of scripts, were conducted with randomly selected residents, shopkeepers and tourist drivers. The interviews with the manager of the PNCV (Chapada dos Veadeiros National Park) and the tourism secretary of Alto Paraíso were already pre-established because they were people who could give us more information on the subject.

CHAPTER 3

TOURISM AND ITS CONCEPTUALISATION OVER TIME

Tourism was first defined, according to The Shorter *Oxforder English Dictionary* (2007), around 1810 as the "theory and practice of travelling for pleasure", while a tourist would be "a person who makes one or more excursions, especially for recreation. Someone who travels for pleasure or culture..." (DIAS and AGUIAR, 2002, p. 22).

According to Ruschmann (1997, p. 13), "the word tourism appeared in the 19th century, but *THE* activity has its roots throughout history". The *Michaelis* dictionary (2016) defines tourism as "travelling for pleasure to places that arouse interest". The Aurélio dictionary (2016) conceptualises it as: "Travel or excursion, made for pleasure, to places that arouse interest."

Fourastié *(apudRUSCHUMANN*, 1979, p. 17) states that "certain forms of tourism have existed since the earliest civilisations, but it was from the 20th century onwards, and more precisely after the Second World War, that the activity evolved, as a consequence of aspects related to business productivity, people's purchasing power and the good resulting from the restoration of peace in the world".

The Berlin School played an important role in conceptualising the word tourism in the period between 1919 and 1938, as interest in the area as a branch of study gained relevance at that time. This school sought scientific backing in the methods "adopted by the sciences of society at the time" (DIAS, 2005, p. 13).

Robert Glucksmann, one of the scholars in the field who came out of this school, defined tourism as follows: "An overcoming of space by people who flock to a place where they have no fixed place of residence" (DIAS, 2005, p.13).

Another scholar, Schwink, from the same school, wrote that tourism is "the movement of people who temporarily leave the place of their permanent residence for any reason related to spirit, body or profession" (DIAS, 2005, p. 14).

Artur Bommann, also an exponent of this school, stated that "tourism is the set of journeys whose purpose is pleasure, commercial or professional reasons or other similar ones, and during which the absence from the habitual residence is of short duration" (DIAS, 2005, p. 14).

However, in 1937, according to Dias (2005, p. 15), the League of Nations[1] , represented by its Statistics Committee, sent a proposal to the Council of that organisation to promote the study of tourism. What came to be considered a tourist journey was the absence of any person for 24 hours or more from their habitual residence.

In this proposal, however, the Committee classified travellers in the following categories as tourists,

---

[1] It was an international organisation, which gave rise to the UN, conceived in 1919 in Versailles, in the suburbs of Paris, where the victorious powers of the First World War met to negotiate a peace agreement.

as described by Dias (2005, p. 15):

> a) people travelling for pleasure or for family, health, etc. reasons;
> b) people attending a meeting or in a service capacity (scientific, administrative, diplomatic, religious, sporting, etc);
> c) business travellers;
> d) visitors to cruise ships, even when their stay is less than 24 hours [...];

Apart from the cases mentioned above, the same Committee decided to exclude the following cases from consideration as tourists, as quoted by Dias (2005, p. 15):

> a) people who arrive with a work contract or not to take up a job in the country or to carry out a professional activity;
> b) people who take up residence in the country;
> c) students and young people staying in hostels or schools;
> d) people who live on the border or people who live in one country and work in another;
> e) travellers in transit without stopping in the country, even when the journey takes more than 24 hours.

In 1945, the United Nations Organisation (UNO), which was the successor to the League of Nations, even though it continued the principles of these concepts adopted previously, added six months to the period for tourists to stay in a region other than their own. However, the UN's merger with other organisations increased this period to no less than twelve months.

Marcel Gautier, as quoted by Dias (2005, p. 16), formulated one of the most synthesised definitions of what would come to be considered tourism, for which "it is the set of economic and social phenomena caused by travelling".

Having considered the first definitions of what tourism is, the need then arose to define what a tourist is. Glucksmann (apud DIAS, 2005, p.15) summarises that "people who, under the legal provisions of a country, are obliged to formalise their residence after a certain period of stay are practically not included in tourist traffic".

At a Congress on International Travel and Tourism in Rome in 1963, organised by the UN and the *International Union of Official Travel Organisations* (IUOTO)[2] , it was agreed that the terms *visitor* and *tourist* would be adopted, the latter being designated as "any person visiting a country other than that of their normal place of residence". (DIAS, 2005, p. 15). While the term *visitor* is used, according to Oliveira (2002, p. 37), to designate a person who visits a place other than that of their habitual residence.

Following the International Conference on Travel and Tourism Statistics held in Ottawa in 1991, the committee responsible for this event approved a report by the World Tourism Organisation (UNWTO) in 1993,

---

[2] International Union of Official Travel Organisations, predecessor of the World Tourism Organisation (UNWTO)

in which tourism was defined as:

> The activities that people carry out while travelling and staying in places other than their usual surroundings, for a consecutive period of less than a year, for leisure, business or other reasons, not related to carrying out a paid activity in the place visited. This definition includes all visitor activities, such as tourists (overnight visitors) and excursionists (day visitors) (DIAS, 2005, p. 18).

This parameter also includes *excursionists*, who are temporary visitors who stay less than 24 hours in the country they are visiting, as stated by the UN at the same Congress.

It took a while for Brazilian tourism to be legalised, according to describes Cerqueira (2009):

> The trajectory of tourism legislation in Brazil began in 1938, when the state set out to control the activity more closely. Getúlio Vargas signed Decree-Law 406, which stipulated that transport companies could only operate and sell air, sea and road tickets with government authorisation. This interventionism and state control lasted for another 53 years, compared to just 17 years of free market.

different types of tourism.

3.1 TOURISM IN BRAZIL FROM 1960 TO 2016

Studies on tourism in Brazil began in the 1960s, more specifically in 1964, with the publication of the Tourist Guide by the now-defunct Tourism and Events Division of the Ministry of Industry and Commerce. After this period, with the country politically reordered and managing the recessionary period from 1962 to 1967, Brazil "experienced a process of accelerated modernisation through authoritarian means, compatible with the dynamics of industrial capitalist accumulation" (TASCHNER, 2003, p. 9). However, it was in 1966 that the Federal Government, giving recognition to this activity, created the National Tourism Council (CNTur) and through Decree-Law 55[3], of 18 November that year, EMBRATUR, at the time the Brazilian Tourism Company, and today called the Brazilian Tourism Institute (IBT), which would have the mission of formulating, coordinating and enforcing the National Tourism Policy.

> According to Decree-Law 60.224/67, Embratur was to study and propose to the CNTur the normative acts necessary to promote the National Tourism Policy, as well as those concerning its operation (CRUZ, 2000, p. 51).

Santos Filho (2005, p. 1) argues that EMBRATUR was created with other objectives, namely to "explicitly coordinate the development of Brazilian tourism. And, implicitly, to remake Brazil's image abroad, so there was nothing better than an organisation through which to publicise the natural beauty of an exotic,

---

[3] This law was repealed on 28 March 1991 by Law No. 8,181

pro-American country [...].

Solha (2002, p. 130) points out that the development of tourism in the country between 1950 and 1969 was the result of a combination of various factors:

- improving equipment and the transport system;
- expansion of communication systems;
- urbanisation and the growth of cities;
- growth of a travelling middle class.

Although it was still in its infancy compared to the speed of world events, this stimulated the beginning of the organisation of the activity in the public and private sectors.

The 1970s, according to Rejowski (1996, p.30), were a very productive decade in the world. with regard to discussions on tourism. The first scientific events in the area began, discussing the reality of Brazilian tourism, the labour market and the needs of the sector, spearheaded by the Brazilian Tourism Congress (CONTUR), the first promoted by the School of Communications and Arts at the University of São Paulo (USP).

In terms of tourism infrastructure, especially for hotels, the 1970s were very productive, even though the pace of growth in the tourism sector was slow. The hotel industry increased the number of beds and improved the quality of its services.

As a result of the various types of long-term financing created at the time, through EMBRATUR and resources from the Financing of Machinery and Equipment (FINAME), for example, as well as tax incentives such as SUDENE and SUDAN, hotel companies doubled their capacity and international companies set up shop in Brazil. The main national chains created during this period were Hotel Nacional Rio, Horsa, Othon, Eldorado and the Tropical de Hotéis chain (DIAS, 1990, p. 41).

According to Solha (2002, p. 131), the first jail was set up in São Paulo in 1975. Hilton, an international hotel chain with 400 flats, a new hotel philosophy and modern management systems. Throughout the decade, other chains were also set up, such as Sheraton (1974), Holiday Inn (1975), Meridien (1975), Novotel (1976) and Club Mediterrané (1977). Thus, there was a period of establishment of several luxury hotels. On the other hand,

> there was the development of alternative means of accommodation, such as campsites, secondary residences and youth hostels. Camping sites developed as a result of the car industry and the expansion of the motorway network, making many Brazilian destinations accessible that had no tourist infrastructure. Secondary residences were consolidated around the major metropolises and hostels since the foundation of the Brazilian Federation of Youth Hostels (SOLHA, 2002, p. 133).

However, it wasn't until 1977 that they issued a document containing the Policy National Tourism Programme, with its basic guidelines, as quoted by SEABRA (2001, p. 17):

<ul>
<li>Publicising and promoting cultural values;</li>
<li>Encouraging internal tourism by building simpler accommodation and reducing the cost of internal travel;</li>
<li>Stimulating tourism from abroad to Brazil.</li>
</ul>

With the growth and diversification of hotel developments, in 1978 EMBRATUR drew up and put into practice the General Regulations for the Classification of Brazilian Accommodation Facilities (SOLHA, 2002, p. 137).

The economic crisis at the beginning of the 1980s and various other obstacles led to the cancellation of long-term financing and tax incentives, leading the segment to economic stagnation, although during this period there were major developments such as the Maksoud Plaza hotels and the Transamérica hotel in São Paulo. Only "air transport continued to grow and, in 1985, Cumbica Airport in São Paulo was inaugurated" (REJOWSKI, 2005, p. 145).

In 1980, in order to attract North American tourists, the Bank of Brazil opened a 50 million dollar credit line for investors in the sector who were interested in expanding their businesses. Two years later, stimulated by EMBRATUR and the Ministry of Labour, the unions organised trips in the low season for workers, following the model adopted by the Social Service of Commerce (SESC), allowing sales workers low-cost accommodation (SEABRA, 2001).

During this period, the railway and road networks were in a precarious phase. However, there was some revitalisation of river transport. An example of this was the reuse of the gaiolas[4] , which for a while were used for tourist trips down small stretches of the São Francisco River in Minas Gerais, or for young adventurers on longer journeys, which could last up to ten days, on the river from Pirapora, MG, to Petrolina, PE (SOLHA, 2002, p. 138).In 1986, given the large flow of tourists and the growth in the number of aircraft, the Brazilian Aeronautical Code (CBA) was created. During this decade, a new type of accommodation emerged in Brazil: flats or apart-hotels, which can be defined as condominiums that offer the basic services of a hotel (LIMA, 1991).

According to Solha (2002, p. 140),

> With the seeds of tourism already planted, companies organised themselves into associations and the government began to realise that tourism is much more than investing in hotels. As a result, tourism began to be seen as a serious and professional activity that doesn't provide immediate solutions to structural problems, especially economic ones.

Among the numerous projects, those planned for the northeastern coastline stand out, such as the Parque das Dunas-Via Costeira Project (Natal-RN), the Cabo Brancos Project (PA), the Costa Dourada Project (PE and AL) and the Linha Verde Project (BA), made possible by the incentives created to promote the hotel

---

[4] Old steamboats built in the USA to navigate the Mississippi, imported in 1917 to navigate the Amazon and transported to the São Francisco River.

sector (Cruz, 2000).

The 1990s was the time when the world globalised. Advances in technology and the arrival of the internet were decisive factors for tourism in Brazil to regain its strength. Large hotel chains established themselves in Brazil and new airlines began operating out of the country's main airports. "With the internet, competition has reached travel agencies and operators, taking over their international sub-sectors" (TASCHNER, 2003, p. 9).

In May 1990, tourism in Brazil and Latin America was discussed at a scientific event organised by the World Association for Professional Training in Tourism (AMFORT), held at USP's School of Communication and Arts. Entitled Turismo: O Grande Desafio dos Anos 90 (Tourism: The Great Challenge of the 1990s), this seminar analysed the main themes of that moment, highlighting issues related to human resource training, development, tourism policy and tourism planning (SOLHA, 2002, p. 141).

In the 1990s, analyses Ferreira (2009, p. 18), tourism continued to be an important source of foreign exchange for developed countries. Apart from Europe, the USA and Canada were the regions that received and sent the most tourists in the world, as they had strong economies and a level of human development that allowed their populations to have more disposable income and more free time to devote to leisure activities such as tourism.

However, the main source markets for tourists to Brazil were the countries that make up the Southern Cone Common Market (MERCOSUR), such as Argentina, Uruguay and Paraguay, followed by European countries (Germany, Italy, France, England) and the USA (EMBRATUR, 1999).

In this period, as described in Observatório do Mundo do Trabalho: Turismo e Hospitalidade (2012, p 19),

> Tourism was placed among the government's priorities. EMBRATUR, until then a public company, became the Brazilian Tourism Institute in 1991, categorised as a special authority. By 1992, guidelines had already been established for the National Tourism Policy (PNT) and its operationalising instrument, the National Tourism Plan (PLANTUR). In 1994, the National Tourism Municipalisation Plan (PNMT) was created [...]. It wasn't until 1996 that EMBRATUR launched the PNT for the three-year period 1996-1999.

For Rejowski (2005, p. 149), the 1990s were also marked by an increase in outbound tourism. Tourism moved away from being a marginal activity and demonstrated that it had the infrastructure, equipment and services with the potential to compete with international tourism. However, while this potential was not being utilised, Brazilian tourists were heading abroad in droves. Outbound tourism only declined towards the end of the decade, when actions were implemented to develop national infrastructure, cheapen airline tickets and improve accommodation, allowing domestic tourism to consolidate.

In 1999, with the drop in outbound tourism, conditions were created for the development of the country's tourism infrastructure. EMBRATUR's programmes to encourage domestic tourism were accompanied by the cheapening of airline tickets and the improvement and diversification of the country's accommodation facilities. This period saw the consolidation of domestic tourism, when trips began to be popularised "through partnerships and charters that made it possible to offer low prices, financing through credit card operators and, above all, the provision of quality equipment and services" (SOLHA, 2002, p. 144).

The 21st century, as Ferreira (2009, p. 19) points out, is being marked by numerous attempts to promote the sustainable development of tourism, "through the responsible management of natural and cultural resources, the conscious use of infrastructure and tourist facilities in order to minimise the negative effects that the development of the activity can cause".

At first, public administrations were not prepared for the increase in demand for tourism that developed in Brazil at that time. This led them, especially the municipal ones, to "authorise developments in areas of water sources, the destruction of important ecosystems, the implantation of allotments near rivers and the destruction of forests, in other words, decisions taken in the light of disordered tourist development" (DIAS, 2003, p. 25).

The most popular forms of tourism in this decade were ecological and ecotourism. According to Ruschmann (1997, p. 27), tourism in green areas is not something temporary, but rather "an awareness of the need to protect the environment, given that the great demand for contact with nature has intensified and has become an important commercial argument".

Ecological tourism, which can also be called nature or green tourism, according to Domiciano and Oliveira (2012, p. 5), is that which takes place through the "travelling of people to natural spaces, with or without reception facilities, motivated by the desire/need to enjoy nature, passively observe the flora, fauna, landscape and scenic aspects of the surroundings".

Beni, quoted by the same authors, goes on to define ecotourism as:

> The form of tourism that takes place in natural spaces delimited and/or protected by the state or controlled by some organisation, such as local associations or non-governmental organisations (NGOs), with planning for the sustainable use of their natural and cultural resources, in which activities such as eco-tourism can be carried out, provided that the restrictions on the use of these spaces are strictly observed.

In 2002, the UN established this year as the International Year of Ecotourism. During this period, the UNWTO and the UN convened the World Ecotourism Meeting in Québec, which established the Québec Declaration, a milestone in recognising tourism's contribution to the achievement of the Millennium Development Goals (MDGs). Particularly in relation to poverty reduction, environmental conservation and the creation of opportunities for women and young people (UNWTO Annual Review, 2009).

In this same context, in 2007, the UNWTO, together with other UN institutions, organised the First International Conference on Climate Change and Tourism. "This meeting was responsible for discussing and presenting measures to tackle and minimise climate change, which is compromising tourism and damaging the world" (FERREIRA, 2009, p. 21).

In 2009, it was predicted that tourism figures would remain flat or even fall slightly. Even so, when compared to other sectors of the economy, such as construction or the automotive industry, for example, tourism figures were more resilient to the troubled economic situation (UNWTO, 2009).

In the 2010s, Brazil was awarded the honour of hosting the world's two biggest sporting events, the 2014 FIFA World Cup and the 2016 Olympic Games. The 12 host cities chosen for the football tournament developed their urban mobility projects and stadiums and hotels multiplied across the country, both in retrofit projects[5] , and in the construction of new developments, seeking to meet the requirements of the International Football Federation (FIFA) and the International Olympic Committee (IOC) in terms of quantity and quality (CALFATI, Hotéis Magazine, 2011).

According to data released by the Brazilian Ministry of Tourism in the Mais Turismo, Mais Desenvolvimento - Indicadores report (Brazil, 2013), Brazilian tourism grew by 6% in 2012, two percentage points above the annual world average. In international tourism, Brazil also gained positions: the business, events and conventions segment was booming and the related services are in areas such as hospitality, gastronomy and hotels.

According to the report, the highlight of the national tourism account has been domestic consumption by families, driven by the growth of the Brazilian middle class and the increase in employment levels in the country. Domestic tourism now accounts for approximately 85 per cent of the sector's revenue in Brazil.

According to the MT, in 2014 around 1 million foreign tourists from 202 countries visited Brazil during the World Cup. The 21 airports that served the 12 host cities had a daily average of 485,000 passengers.

According to the National Tourism Council (CNTur) (2011-2014, p. 57), preparing for the mega-events that have taken place in this decade is both a challenge and an opportunity, not only to consolidate and recognise tourism as an important socio-economic development factor for the country, "but also to build a new level of quality for territories and the network of cities in Brazil, particularly with regard to accessibility and urban mobility".

In 2016, according to the President of Embratur, Vinícius Lummertz, "from these Olympic Games onwards, we're going to experience an important cycle for the country, which will be integrated into tourism opportunities for the whole world due to its own way of organising events".

According to a report released by Rio de Janeiro City Hall, the city received 1.17 million visitors during the tournament, which took place between 5 and 21 August 2016. Of these, 410,000 were foreign

---

[5] A trend in architecture and design that emerged in Europe in order to solve a problem: what to do with such a large number of old and historic buildings that are unusable, or with outdated technologies that make it impossible to use them? This trend has emerged as a way of revitalising buildings and other constructions by bringing them new technologies and more promising designs.

tourists. Also according to the City Hall, Rio de Janeiro received 243,000 tourists during the Paralympics, which took place between 7 and 18 September of the same year.

CHAPTER 4

PROTECTED AREAS AND CONSERVATION UNITS

The management of natural areas, over time and given the intensity of environmental problems, has evolved both in terms of its concept and its importance.

According to MILANO (1997), since the 19th century, environmentally conscious people and institutions have sought to protect natural areas in order to maintain samples of biological communities and natural beauty.

According to Hauff (1997 apud MILANO, 2004, p.5),

> In addition to protecting scenic beauty for future generations, the conservation of these areas covers objectives ranging from maintaining natural diversity at all levels, favouring scientific research, providing environmental education and recreation, protecting historical and/or cultural sites, managing forest and wildlife resources, ensuring environmental quality and regional economic growth, offering technological flexibility and defending investments from environmental dilapidation.

The denomination of Conservation Units given by the National System of Conservation Units (SNUC), Law No. 9.985, of 18 July 2000, to natural areas that can be protected due to their special characteristics, states that Conservation Units are:

> Territorial spaces and their environmental resources, including jurisdictional waters, with relevant natural characteristics, legally instituted by the Public Power, with conservation objectives and defined limits, under a special administration regime, to which adequate guarantees of protection of the law apply (art. 1, §I).

Law 9.985/2000 groups protected areas into two categories, which are integral protection, when they aim to preserve nature and only allow indirect use of its resources, namely: ecological stations, biological reserves, national parks (as shown in Figure 2), natural monuments and wildlife refuges. And sustainable use units, which are those that aim to make nature conservation compatible with the direct sustainable use of part of its natural resources, such as: environmental protection areas, areas of relevant ecological interest, national forests, extractive reserves and fauna reserves (HAUFF, 2004, p.6).

According to the Pro-Nature Foundation (FUNATURA, 1991), the first group aims to preserve natural processes and genetic diversity with as little human interference as possible, allowing only indirect use of resources. The second group seeks to reconcile, as far as possible, the preservation of genetic diversity and natural resources with the direct, moderate and sustainable use of some of these resources, limiting anthropogenic alteration to a level compatible with the permanent survival of plant and animal communities.

According to Milano (1991, p.19), the units located in populated territories have become "islands" "isolated and surrounded by anthropised systems which, when they don't produce strong pressures, at the very least make certain conservation objectives partially unfeasible". Oldfield (1988, apud HAUFF, 2004, p. 7), on the other hand, states that legal protection has often not been enough to maintain the integrity of protected areas, even when there are management activities in place.

Hauff (2004, p. 8) argues that normally the inhabitants of the regions that are considered protected areas see these spaces as government impositions restricting their traditional rights, in which the social systems for protecting their communities are not included. As a result, agriculture invades protected areas and forest products are easily exploited within their boundaries, laws are not obeyed, control of these areas fails and the unit is gradually destroyed.

a) THE LANDSCAPE IN CONSERVATION UNITS

The creation of protected natural areas began with the demarcation of Yellowstone National Park in the United States in 1872. With the start of the Industrial Revolution and, consequently, population growth and changes in the urban environment, "life in the countryside became the ideal of social classes not involved in agricultural production, taking nature as a place for contemplation, reflection and spiritual isolation". (SILVA, 2003, p. 33)

Landscape didn't come into existence after the birth of man, it was already there. But it's only when man pays attention to the landscape that its concept emerges. Landscape is what you see. What is real, what is experienced, what is felt is different for each human being. They make personal selections and value judgements according to their individual analysis of perception. This analysis is influenced by social, cultural, environmental and emotional factors, depending on the type of landscape used by each person (BOLSON, 2004). According to Gomes (2001, p. 36):

> Landscape as representation results from the apprehension of the individual's gaze, which in turn is conditioned by physiological, psychological, sociocultural and economic filters and the sphere of recollection and recurrent memory of the paths travelled by tourists.

Another author who touches on the subject is Rodrigues (2005, p. 50), who suggests that "the landscape of a place can be a very valuable tourist resource, because it determines whether a place is more or less touristy, or whether a place is touristy or not".

Given the importance of informing tourists about the various forms of tourism on offer according to the characteristics of each region or place, the conceptualisation of the types of tourism that exist in the world was created. These concepts are also of great value to those wishing to invest in the sector. Let's take a look at the definitions of ecological tourism and ecotourism, according to OLIVEIRA (2002, p.77), which will be the fruit of this study and which are in the following categories:

4.1.1 Alternative tourism

Concern about the environment began in the 1970s in central capitalist countries. In Brazil, this happened at the end of the 1980s and beginning of the 1990s, "with the expansion of means of transport and a total transformation of human life in large urban centres" (DIAS, 2003, p. 24).

A new type of tourism began to be practised at the end of the 1980s, given the worldwide ecological awareness and the great demand for environments considered healthy and surrounded by nature. "The improvement in quality of life and a constant change in habits and values gave birth to what we now call alternative tourism" (DIAS, 2003, p. 16).

According to Magalhães (2002, p.28),

> alternative tourism was developed in Europe with the aim of satisfying the needs of a clientele with aspirations and motivations arising from a new contemporary reality, as well as trying to meet the demands of the environment.

This search for contact with nature and the ecological concern of both developed and developing countries, with the valorisation of natural environments by high-income groups, has led to an increase in demand for this type of tourism. This has created a contrast with what happened in the daily lives of these people who lived in large centres and led stressful lives, and who had a landscape there that was - for the most part - poorly preserved or poor.

The result of this search was an increase in demand for services that met their minimum needs, "such as tranquillity, contact with nature rich in colour and animal life and a healthy environment" (DIAS, 2003, p.16).

Tourism scholars Wearing and Neil (2001, p. 2) explain the subject as follows:

> The term alternative implies its opposite. Thus, alternative tourism is the opposite of what is seen as negative or harmful in conventional tourism: it is characterised by the attempt to minimise the visible negative environmental and socio-cultural impact of people on holiday by promoting radically different approaches to conventional tourism.

The so-called environmental crisis that has been transforming contemporary society plays a fundamental role when it comes to choosing a tourist route.

With the growing contamination of the soil, atmosphere and water, "the destruction of the ozone layer and the increasing loss of species of fauna and flora, the search for routes that preserve the environment and value nature has only increased" (DIAS, 2003, p. 18).

4.1.2 Ecological tourism and ecotourism

As Dias (2003, p. 19) points out, ecological tourism is practised by people who enjoy direct contact with nature in places that are no longer part of the big city scene. Regions where nature remains intact or whose landscapes have been naturally conserved are the most popular. Travellers want to breathe fresh air and observe the beauty of their surroundings. They do this in a variety of ways: hiking on trails through woods and forests, riding animals, travelling in environmentally friendly equipment, diving in clear waters, climbing mountains, whale watching, etc.

Ecotourism, according to EMBRATUR (1994, p. 19),

> is a segment of tourism that makes sustainable use of natural and cultural heritage, encourages its conservation and seeks to raise environmental awareness through the interpretation of the environment, promoting the well-being of the populations involved.

According to Rodrigues (1992, p. 23), ecotourism is an activity in which the main purpose of leisure travel is to use and enjoy the natural resources existing in the area.

chosen destination". The same author mentions the necessary conditions for the development of ecotourism:

> the protection of the environment with a view to its conservation, the promotion of benefits for local communities, with respect for their culture, the expansion of knowledge about nature and the ethics that ecotourists must have in order to conserve the environment.

Rodrigues (1999, p. 39) adds that ecotourism,

> One of the most recent forms of tourism - alternative tourism - is characterised by trips to relatively undisturbed and uncontaminated nature reserves, with the specific aim of studying, admiring and enjoying the landscape, fauna and flora, as well as integrating travellers with their surroundings and, in particular, with the local communities.

For the Ministry of the Environment (2001),

> In Brazil, ecotourism is seen by both the government and scholars as an economic alternative with a socio-environmental sustainability profile and as a means of conserving natural, cultural and historical resources, as well as generating benefits for local communities, such as income and employment.

Nowadays, the tourism market in many countries is taking environmental problems into account. Given the extent to which the natural features of the environment have been eroded, tourists avoid visiting affected and eroded areas. However, natural environments and their elements have become a great attraction for exploration, education and the spirit of adventure, thus giving rise to a new market, which is the practice of both ecological tourism and ecotourism. Thus, there is a renewal of tourism, whose practitioners seek calm, adventure and a wealth of knowledge about the regions visited.

CHAPTER 5

CONCEPTS AND REPRESENTATIONS OF PARKWAYS

The concept that permeates the various definitions is that Park Roads have this name because they are not ordinary roads, i.e. they are roads characterised by their unique aspects in archaeological, cultural and historical terms, as well as having other features such as natural and recreational areas and high scenic quality (NATIONAL SCENIC BYWAYS PROGRAM, 2009).

Considering that, with the creation of the Park Roads, tourist attractions have significant landscape value, the relationship between the study of landscape and tourism is evident, as is the need to preserve natural and humanised landscapes as tourist attractions. "This preservation is also directly related to the issue of protected areas, which should, in the case of the SNUC, consider Park Roads as a category of their own, providing for their tourist exploitation among their management objectives" (PIRES and TIAGOR, 2010, p. 2).

It is extremely important to consider that even given what the Conservation Units represent, Brazil only has 1.87 per cent of its territory fully protected by them. "This is considering that there is variation from state to state" (SORIANO, 2006, p.4).

According to the Brazilian Environmental Institute (IBAMA, 1989), these areas were insufficient in terms of land area and did not represent the different ecosystems adequately, which resulted in damage to the biota, as well as administrative and inspection difficulties due to a lack of resources and a defined nomenclature for management categories.

In Brazil, as in the United States, there is a dearth of literature on these roads. The first official reference to park roads in Brazil dates from 1982 and is part of the Plan for the Brazilian Conservation Units System, Stage II, according to the Brazilian Forestry Development Institute (IBDF) and the Brazilian Foundation for Nature Conservation (FBCN), also from the same year.

In 2009, the so-called parkways were regulated by the US National Park Service, which describes them as follows:

> These are roads built in such a way as to protect and make it possible for people to enjoy the scenic beauty and interesting historical points along the route. A particular aim of these roads is to prevent the erection of signs, billboards and other information or advertising that could detract from the natural beauty along the road. (NATIONAL PARK SERVICE, 2009)

Below are the concepts of Park Road found in Brazilian literature specialising in the area.

Figure 1 - Table showing the Parkway concepts in Brazil

	Document - Publication	Concept - Definition
1	BRAZILIAN INSTITUTE OF FORESTRY DEVELOPMENT - IBDF / BRAZILIAN FOUNDATION FOR NATURE CONSERVATION - FBCN	It is a linear park that includes all or part of roads of high panoramic, cultural or recreational value. The boundaries are established in such a way as to include the land adjacent to both sides of the road, in order to protect the integrity of the

		panorama, related resources and recreational and educational activities.
2	SILVA, Lauro Leal	It is a linear park of high educational, cultural, recreational and panoramic value that protects strips of land along stretches or the entirety of paths, roads or access routes, and whose boundaries are established with a view to protecting their characteristics and maintained in a natural or semi-natural state, avoiding works that disfigure the environment.
3	BARRO S, Lídia Almeida	A management category whose main objective is to protect, in part or in whole, roads with banks of great natural, semi-natural or cultural scenic beauty. The areas adjacent to these linear parks can be public or private.
4	S.O.S ATLANTIC FOREST FOUNDATION	The Park Road is a permanent museum route that crosses Conservation Units or areas of significant environmental and landscape interest, set up with the aim of combining environmental preservation with the sustainable development of the region, by promoting ecotourism and environmental education, leisure and cultural activities.

Source: SORIANO (2006)

a) CREATION OF PARKING ROADS IN BRAZIL

It was during the World Wars, around 1930, after the New York stock market crash of 1929, that the first parkways were created in the United States. The Blue Ridge Parkway was the first to be built, proposed by the

then US President Franklin Roosevelt, in what was called the New Deal Work Programme. The purpose of this road was to create jobs, since construction played an important role in enabling people living in the mountains to earn some form of income (PARKWAY PARTNER ORGANISATIONS, 2009).

Located in Germany, between the regions of Bavaria and Baden-Wurttemberg, and running from Wurzburg to Fussen, the Romantic Road was so named by travel agents in the early 1950s. It stretches for 350 kilometres and along the two regions there are several medieval tourist towns, as well as castles and local shops. This road is not a motorway, but a scenic route for passenger vehicles.

There is a cycle path along its entire length, as well as special tourist packages for cyclists. However, there is an alternative route for freight transport or rapid transit. There is also the Europa Bus Romantic Road, which connects the towns on the tourist route in both directions of the road. (Romantic Road Germany, 2010)

The aim of the parkways, unlike traditional roads, was not to provide the quickest and shortest route, but a moderate drive to enable contemplation of the landscape (NATIONAL PARK SERVICE, 2009). The creation of these roads differed from conventional roads in at least eight respects, which were mentioned as early as 1938, as we see below:

1) They are designed for non-commercial and recreational use;
2) Avoid unwanted buildings and other constructions on the sides of roads that could deteriorate the original road landscape;
3) Built with a wider right of way to provide an insulating strip of land between the road and private properties;
4) Eliminate façades and access rights to preserve natural scenic values;
5) Preferably in a new location, bypassing communities and avoiding traffic jams;
6) Aimed at making the best scenery it crosses more accessible, so the shortest or most direct routes are not necessarily a primary consideration;
7) Eliminate major level crossings;
8) Have entrance and exit spaces at distant intervals to reduce interruptions to the main traffic (NATIONAL PARK SERVICE, 2009).

According to Santos and Martins (2008), in Brazil the creation and management of a Park Road must take place within one of the categories of the law that establishes the SNUC, Law No. 9.985 of 2000.

BRASIL

Figure 2 - Existing **parkways** in Brazil **Source:** SORIANO (2006)

Some Park Roads proposed in Brazil are included within pre-existing PAs, so they must respect the limitations imposed by legislation on the corresponding category. Others fall within areas whose protection has not previously been designated by public act, so their creation and regulation must define which category of UC they will fall within.

In view of this, a comparative study carried out by Tricário, Oliveira, Rossini and Carvalho (2012) to find definitions for park roads in the country shows some of the main roads considered to be park roads in Brazil.

5.1.1 State highways MS-184 and MS-228 (Pantanal, MS, Brazil)

The MS-184 and MS-228 state highways (Figure 3) have been considered a PA by the Mato Grosso do Sul government since 1993.

Its routes go back to the original routes of the great cattle drovers and the natural paths of the Pantanal.

Figure 3: State highways MS-184 (left) and MS-228 (right), Pantanal, MS **Source:**
https://www.google.com.br/search?q=Rodovia+MS+184+photos

These roads were only built with embankments to ensure travel at any time of year. Along the route, the hotel services sector predominates with inns and farm hotels. In the Pantanal flora, common species bloom at different times of the year.

Adventure sports are widely practised and there is a constant presence of images that appeal to nature, which emphasises the tourist vocation of the area. Sport fishing is also a constant practice and there is equipment to enjoy the fauna and flora. On the road, there are 70 wooden bridges for observing birds and animals in large numbers. There is also undergrowth with thousands of streams and natural lakes (ESTRADA-PARQUE PANTANAL, 2010).

5.1.2 Serra da Cantareira Park Road (São Paulo, Brazil)

One of the largest native urban forests in the world, the Serra da Cantareira State Park (Figure 4) covers 8,000 hectares. In 1994, this region was recognised by the United Nations Educational, Scientific and Cultural Organization (UNESCO) as part of the São Paulo Green Belt Biosphere Reserve.

Access to the park is via the road, which, despite the large influx of tourists at weekends, has no cycle lanes, public pavements or cycle paths. Public transport is the most widely used means of transport, as well as passenger cars. Information plaques about the localities and sectors, as well as relevant geographical aspects, are dotted around the park.

Figure 4: Serra da Cantareira Park Road
Source: https://www.google.com.br/search?q=fotos+Estrada-Parque+da+serra+da+cantareira

The entire road, which makes it easier to locate the existing routes. As on the Serra da Mantiqueira road in the state of São Paulo, Atlantic Forest vegetation predominates, and the native vegetation has been replaced by undergrowth in order to make the road look more attractive and interesting.

5.1.3 Itacaré Park Road (Bahia, Brazil)

Completed in 1998, the Itacaré/BA Park Road (Figure 5) was the first motorway in Brazil to have environmental monitoring throughout its construction phase. It is 65 kilometres long and has several beaches along the way, as well as trails, waterfalls, inns, places for tree climbing, extreme sports, etc. The tourism practised is sun and sea, with a strong appeal to a tropical and luxurious nature. The means of transport used is public transport, linking the cities of Ilhéus and Itacaré. There is also a disorganised flow of tourists due to the lack of structure and property speculation, which worries local residents.

The Atlantic Forest vegetation predominates along the road and there are only two places considered to be viewpoints for contemplating the landscape. The signposting is at odds with what the region has to offer, as the standards differ from the visual communication. With large areas of environmental preservation, the fauna is very rich and there are a number of attractive birds for tourists to enjoy (Estrada-Parque de Itacaré, 2011).

Figure 5: Itacaré Park Road. Bahia **Source:** https://www.google.com.br/search?q=Estrada-Parque+of+itacaré+photos

5.1.4 Park Road RJ-163 and RJ-151 (Visconde de Mauá / Capelinha - RJ)

The paving of this road is considered the main infrastructure project of the Rio de Janeiro Tourism Development Programme (PRODETUR-RJ). However, by 2014 only the paving had been completed. This is a road that includes the towns of Capelinha and Visconde de Mauá (RJ-163, Figure 6 left) and the town of Maromba, in Itatiaia, and Ponte dos Cachorros, in Resende (RJ-151, Figure 6 right). The Mauá Park Road is located within the Serra da Mantiqueira APA (MAIA, 2014, p. 26).

Figure 6: RJ-163 (left) and RJ-151 (right) park roads **Source:** https://www.google.com.br/search?q=RJ+163+photos

According to Maia (2014, p. 31), the initial project for the Visconde de Mauá Park Road included not only paving but also the following features:

> - Implementation of aerial and underground zoopassages on certain stretches of the RJ-163 motorway;

> *c* Construction of wooden decks at the viewpoints of the Paraíba do Sul River and the Preto River;
> - Construction of a portal near the town of Capelinha (Resende) - where there would be control of load capacity and the collection of an environmental fee for the conservation of the road;
> - Route operated with traffic restrictions;
> - Access control, with the installation of a gantry on the RJ-163;
> - Drainage system suitable for the region;
> - Signposting with tourist, safety and environmental education information.

The total length of this road is 32 kilometres. As for the initial project objectives described above, only the paving had been completed by 2014 (MAIA, 2014, p. 31).

However, the biggest problem is the demand from the residents of this region,

> who complain that the road has no shoulder and that they are forced to take their chances on foot in the winding bends between the cars that now pass at high speed, despite the maximum allowed being 40 kilometres per hour. Cyclists also complain about the lack of a cycle path and assure us that both the shoulder and the cycle path were provided for in the original project (O Dia, 29/65/2015 edition).

5.1.5 Graciosa Park Road (PR-410)

This road (Figure 7) is considered to be the best equipped Park Road in Brazil, as it has leisure facilities such as kiosks, barbecue pits, car parks, viewpoints and toilets, but it is not managed as a PA. According to the Paraná Environmental Institute, there is no intention of declaring it a Park Road, since its landscapes are within other PAs, the Morumbi Special Areas of Tourist Interest (AEIT) and the Pico do Morumbi State Park (SORIANO, 2014, p. 93).

Figura 7:   Graciosa Park Road, Curitiba, PR
Source: https://www.google.com.br/search?q=estrada+da+graciosa+fotos

According to the author,

> This road has historical attributes that are just as important as its natural ones, given that its original route took advantage of the Indian trails that passed through the plateau and the coast. It was also used intensively by the Jesuits and later served as a route for the tea cycle in Paraná, as it was, for a long time, the only link between the coast and the plateau (SORIANO, 2014, p. 94).

CHAPTER 6

ALTO PARAÍSO DE GOIÁS

The seat of the municipality of Alto Paraíso de Goiás (Figure 8) is located at the junction of the GO-118 and GO-327 motorways, in the Chapada dos Veadeiros region. Its territory includes 40% of the PNCV area, which corresponds to 10% of the municipality's area (PEREIRA, 2000, p. 18). Known for its mysticism, people believe that the land in this region is protected by extraterrestrial beings, which prevents catastrophes from happening there.

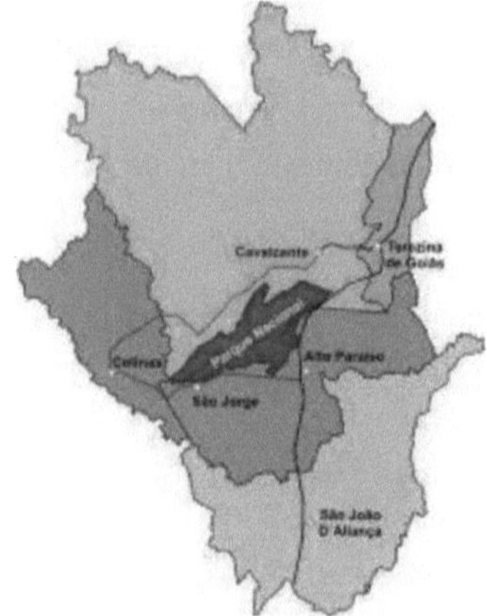

Figura 8: Map covering the region of Alto Paraíso, São Jorge and PNCV **Source**: www.revistaturismo.com.br/Dicasdeviagem/altoparaiso.htm

This municipality arose around 1750, close to the sources of the São Bartolomeu River, where a small colonisation nucleus was established, formed by a group of farms, growing coffee and raising cattle, in the region known as Veadeiros, situated on the edge of the road that linked the south of the state of Goiás to the regions of Cavalcante and Palmas (ALBUQUERQUE, 1998; BEHR, 2000).

The first residents of Alto Paraíso de Goiás were the Indians, followed by the bandeirantes who came in search of minerals to exploit the area. According to the Alto Paraíso City Hall Portal (2016),

> The first records of human occupation in the region are of indigenous tribes such as the Cayapós, the Xavantes and the Guayazes. Then came the bandeirantes in search of gold mines and escaped slaves, starting the mining cycle around the Chapada dos Veadeiros region, which led to the emergence of Cavalcante in 1740.

Alto Paraíso was originally called Veadeiros and belonged to the municipality of Cavalcante. According to the Town Hall website, the site was a farm founded by Francisco de Almeida, from where the first village of residents sprang up. Its emancipation came about through political intervention by the state of

Goiás in 1953, and it still goes by the name of Veadeiros.

In 1963, Alto Paraíso was given its name by a vote of councillors. It wasn't until 1981 that construction began on the town hall and the project that would turn the town into a tourist centre.

Mining was the main reason for the region's settlement. Today, Alto Paraíso de Goiás lives off the income from local tourism, which is geared towards nature and is an alternative form of tourism. The locals are particularly respectful of nature and live peacefully side by side with it.

The town has a number of hotels and restaurants that leave nothing to be desired. Many of the town's businesses are owned by people from other places, mainly Brasília/DF, who go there in search of the calm of the countryside and a source of income.

a) DISTRICT OF SÃO JORGE

Vila de São Jorge (Figure 9) is a district of Alto Paraíso de Goiás, located 260 kilometres north of Brasília, 460 kilometres from Goiânia and 36 kilometres from the municipal seat, a small village in the interior of Goiás. The village has approximately 500 inhabitants. (ECO.TUR.BR, 2016) The community, which has the main gateway to the PNCV, has had its economy almost entirely turned over to tourism. "São Jorge is the gateway to the Chapada dos Veadeiros National Park, an old mining village from the early 20th century, whose main activity today is ecotourism, an ecological and sustainable alternative for the region's residents." (MAYOR OF ALTO PARAÍSO DE GOIÁS, 2016).

Figure 9: Vila de São Jorge seen from above **Source:** www.revistaturismo.com.br/Dicasdeviagem
The map below shows the location of Vila São Jorge and all the towns that surround it.

Figure 10: Map covering the region of Alto Paraíso de Goiás, São Jorge and the PNCV **Source:** https://www.google.com.br/search?q=Fotos+mapas+sãojorge

Crystal mining was the main reason why people from different parts of Brazil, especially Bahia, came to the north of the state of Goiás at the beginning of the 20th century.

The fact that the value of rock crystal was manipulated by the international market, since Brazil was almost exclusively the only producer of natural crystal, meant that the price varied constantly. At the start of the Second World War, the war industry created the first great demand for the ore to manufacture ammunition, which also served as components for communication devices (SILVEIRA, 1997, p. 10).

From 1930 to 1948, crystal was well-accepted but suffered price changes. From 1948 to 1949, crystal lost its value and mining activity weakened. At this time, the prospectors made a great discovery in the lower Garimpão region, which was unpopulated and surrounded by mountains.

At the same time that crystal prices were falling, due to low demand and low prices, in 1949 there was a fire in the only warehouse in Garimpão, causing the owner to move to the area that is now the district centre, and there he sold the crystal in a small shed, thus giving birth to Baixa da Chapada dos Veadeiros, which became the garimpeiros' headquarters and later became the town of São Jorge.

According to Silveira (1997, p. 10):

> As they say, the Chapada has experienced other periods of "mining influence", in other words, times when the activity expanded. One of the most cited is 1952, which may have involved almost 3,000 prospectors in the region. This new demand and consequent increase in prices can be explained by the start of the Korean War in 1950.

São Jorge was made a district of Alto Paraíso by Law 499/96 of 6th December 1996. There are some accounts of the place's name. According to Silveira (1997, p. 8), "Father Beno Bakermans, a Dutchman who has lived in Chapada since 1958, says that the prospectors got an image of São Jorge and on the initiative of Severiano da Silva Pires, the town was baptised with this name".

The crystal lost its value again in 1956, creating a crisis for the prospectors. Many returned to their homelands, others went to work building Brasilia. A few remained in São Jorge (OLIVEIRA, 2002, p.37).

The vast majority of the locals were born there and live in a humble and peaceful way in the village. Most of the locals have taken a course as visitor guides and use this form of service for their family income. Silva (2005) recounts this passage

> The population began to take part in the village's tourist activities, realising the need to create an Association of Local Guides, as well as a Non-Governmental Organisation of Environmentalists.

The village has a municipal school that runs from kindergarten to primary school. Secondary school is in the city of Alto Paraíso de Goiás. The village has a Tourist Service Centre (CAT), which is open every day during business hours. The population has piped water and also artesian wells made in their homes. The village has no tarmac, a preference of most residents, who prefer it to be simpler and to retain its original characteristics.

The locals have built inns and restaurants for tourists, always trying to maintain the originality of the place. According to the Alto Paraíso City Hall Portal (2016),

> Data from the local government (2003) show that the village has 03 restaurants, 01 supermarket, 11 campsites, 18 inns/hotels, 09 snack bars/pizzerias/bars/kiosks, 02 souvenir shops, a craft market, 01 tourist information centre, 01 hospital service centre, water and sewage services, electricity, telephone services, transport services and a tourist guide service.

Nowadays, the small village receives a large influx of visitors, who stay in its inns and campsites. The district has become a major tourist destination and receives more than 2,300 holiday visitors every year. According to Silva (2005), the village of São Jorge receives the flow of ecotourists from the Chapada dos Veadeiros National Park (PNCV) and the attractions in its surroundings, which is also a major tourist attraction.

b) CHAPADA DOS VEADEIROS NATIONAL PARK (PNCV)

Made up of the municipalities of Alto Paraíso de Goiás and Cavalcante, the PNCV (Figure 11) covers an area of 65,514 hectares of high-altitude savannah, of which approximately 60 % is in Cavalcante and the remaining 40 % in Alto Paraíso de Goiás.

Figure 11: 100 and 120 metre jumps in the PNCV **Source:** https://www.tripadvisor.com.br

> The eastern and southern parts face the municipality of Alto Paraíso de Goiás and the western and northern borders are delimited by the municipality of Cavalcante. Two other municipalities, Colinas do Sul and Teresina de Goiás, have parts of their municipalities belonging to the PNCV Buffer Zone (DOMICIANO, 2014, p. 25).

Together, these municipalities concentrate a small population in the state of Goiás. This region's economy is based on subsistence agriculture and extensive cattle breeding.

The Municipal Human Development Index (MHDI) is among the lowest in the world.

State (IPEA/PNUD, 2006).

> The Chapada dos Veadeiros is considered the magnetic heart of the country. It is located in the north-east of Goiás, at an altitude of around 1,700 metres, 230 kilometres from Brasília, between the municipalities of São João d'Aliança, Alto Paraíso de Goiás, Teresina de Goiás, Cavalcante and Colinas do Sul, which make up a plateau of almost 5,000 square kilometres (PORTAL DA PREFEITURA DE ALTO PARAÍSO DE GOIÁS, 2016).

In 1960 there was a proposal to create a Conservation Unit in the region of

However, it was in 1961 that the Tocantins National Park, with an area of approximately 625,000 hectares, was created by Decree No. 49,875, issued by President Juscelino Kubitschek. Articles 1 and 2 of this decree established the creation of the National Park (PARNA):

> Art. 1 - The Tocantins National Park is hereby created in the state of Goiás, in the Chapada dos Veadeiros, subordinated to the National Parks and Forests Section of the Forestry Service of the Ministry of Agriculture.

> Art. 2 - The boundaries of the Park hereby created begin on the right bank of the Tocantins River at the confluence of the Tocantizinho River, following it to its source; thence through the slopes skirting the town of Veadeiros[1] to the source of the Preto River; thence following the same slope to the source of the Santa Rita Stream; thence along the said stream to the confluence with the São Félix Stream; thence along the said São Félix Stream to its confluence with the Tocantins

River;

The name of the park and the region where it is located is a reference to the red deer, which inhabited the area in large numbers in the past (SILVEIRA, 1997, p. 12). Visits to the park should be restricted, but permitted when they are in accordance with the area's management plan and with authorisation from the responsible body. For the National System of Conservation Units (SNUC), a Conservation Unit is defined as:

> Territorial space and its environmental resources, including jurisdictional waters, with relevant natural characteristics, legally established by the Government, with conservation objectives and defined limits, under a special regime, to which adequate protection guarantees apply (SNUC, 2000).

It is quite probable that the main reason for the creation of this Conservation Unit was to preserve nature in the region close to the Federal Capital, since this act was suggested by the "Coimbra Bueno Foundation for the New Capital of Brazil". (SILVEIRA, 1997, P. 11). According to IBAMA (2007), this Conservation Unit was created to protect the water sources of the region.

It is a natural asylum for micro-organisms and various species of flora and fauna. The Park's springs are the starting point for the Tocantins Hydrographic Basins. Silva (2005, p. 11) says that

> The preponderant factor for the implementation and rise of tourism in São Jorge took place in 1961 with the creation of the Tocantins National Park by President Juscelino Kubitschek, with a territory of 625,000 hectares, encompassing an already consolidated town. The park was later significantly reduced by decree 86,596 of 17 November 1981, leaving the area at 65,000 hectares. As a result, São Jorge remained outside the Park, but was destined to maintain indissoluble ties with the federal conservation unit, now called Chapada dos Veadeiros National Park.

The PNCV covers an area of 65,514 hectares of high-altitude savannah. According to the Chico Mendes Institute for Biodiversity Conservation (ICMBio), the park protects animals, plants, trees and many other creatures that live in its reserves. According to Silva (2005, p. 12),

> With a landscape made up of 280 waterfalls and 220 trails and rocky escarpments, the tourist flow in São Jorge has increased by 700% in the last 10 years, receiving around 30,000 people annually (local government data, 2003). As a result, the socio-economic impacts are visible, but still ignored by government policies.

The park has four trails to reach the reserve's natural attractions:

Figure 12: Level of difficulty, duration, signposting and length of trails

Trails	Level of difficulty	Estimated duration	Signalling	Extension
Crossing the Seven Falls	Very heavy	2 or 3 days	Seven oranges	23.5 kilometres
Jumps Trail	Heavy jumping Moderate	4 to 6 hours	Yellow arrows	11 km round trip

| Canyons Trail | Upper moderate | 4 to 6 hours | Red arrows | 12 km round trip |
| Seriema Trail | Very light | 1 h 30 | Blue arrows | 800 metres round trip |

Source: PNCV Administration

CHAPTER 7

GO-239 - Prefeito Divaldo Rinco Park Road

Considered one of the 11 most incredible roads in Brazil, as published on the 4 Rodas magazine website in January 2016, the stretch of the GO-239 Highway (Figure 13), which connects the municipality of Alto Paraíso de Goiás to the municipality of Colinas do Sul, passing through the district of São Jorge, was named the Sulivan Silvestre Park Road, in accordance with State Law No. 13,467 of 30 July 1999, sanctioned by the then Governor of the State of Goiás Marconi Perillo.

Figure 13: Signpost towards São Jorge/Alto Paraíso de Goiás **Source**: The authors (2017)

The name was chosen in honour of the president of the National Indian Foundation (FUNAI), who died after a twin-engine plane crashed near Santa Genoveva Airport in Goiânia, GO, in February 1999.

However, on 28 September 2015, this law was amended by Law No. 19,015, also sanctioned by the same governor. In honour of the mayor of Alto Paraíso de Goiás, who was murdered in 2010, the GO-239 was renamed the Prefeito Divaldo Rinco Park Road.

According to the Environmental Impact Report (RIMA), the road being paved was designated by an act of the Legislative Assembly as

> The Park Road has two distinct segments of engineering solutions for each sub-stretch, due to their different characteristics. The first section begins in the urban perimeter of Alto Paraíso and ends in the district of São Jorge, the second from São Jorge to Colinas do Sul (CONFLORA, 2004, p.6).

According to the Environmental Impact Study (EIA/RIMA) for the road (1998, pg. 1-2), the proposal was due:

In order to define a project that would meet the interests of the population involved and at the same time be compatible with the geo-environmental susceptibility of this stretch, we opted for a Park Road project with a cycle path and tourist and support structures adjacent to the carriageway and lane. In this way, the road will have the characteristics of a Park Road, offering users the option of stopping at points to contemplate the interesting local scenery, especially on the route between Alto Paraíso and the São Jorge District.

Built by the Goiana Transport and Works Agency (AGETOP) at a cost of

In the 36 kilometres of this stretch (Alto Paraíso de Goiás to São Jorge), R$26.3 million was spent on earthworks, asphalt paving, surface drainage, vertical and horizontal signage and complementary works through the Rodovida Construction programme.

The purpose of tarmacking this road, at first, was to facilitate access to the PNCV, since while the road was dirt, in rainy seasons, especially during the winter, the route became almost impassable, given the runoff from the ravines, which cut through most of the stretch, and the road was also quite bumpy.

Domiciano and Oliveira (2012, p. 191) point out that "for tourists, the most significant impacts are the following

highlighted were the precarious conditions of the GO-239 road, with its potholes and dust, which

hinder access to the District/PNCV and its surroundings, with its attractions".

At first, 22 kilometres were paved, which means that 14 kilometres were left behind, as there was some resistance from local residents and environmentalists, who have always been divided on the issue. For them, the tarmac would soon lead to greater environmental degradation of the PNCV area.

However, meetings were held between the local community and the authorities to reach a proposal that would satisfy both the Alto Paraíso de Goiás City Council and the residents of Vila de São Jorge, since the money had already been released in full and it was necessary to complete the work.

According to Lacerda et al. (2004, p. 83), "it is clear that this work will make life easier for the residents of the village, however, the dirt road served as a limitation to tourist numbers, which is common in ecotourism areas".

This stretch was inaugurated on 23/07/2015, a month before the expiry of the licence granted by the Goiás State Department of the Environment and Water Resources (SEMARH), process 19224/2011, Licence 1966/2013, for AGETOP to complete the work.

In addition to the benefits mentioned above, a cycle path was built on this road to provide more safety for cyclists venturing into the PNCV region, with horizontal signage and reflective studs (Figure 14). This cycle path is located only on the right-hand side of the road, towards Alto Paraíso de Goiás/São Jorge, and is divided into two lanes. However, it is not the best way to avoid accidents, as the tarmac is considered narrow.

Figure 14: View of the cycle path on GO-239
Source: The authors (2017)

Despite being an incentive for tourism, this asphalting has brought problems for the community of São Jorge. The main problem was the change in the type of tourists and the increase in their flow, "who now only stay in the village for a weekend or just spend the day there, which is detrimental to the general opinion of the residents, most of whom are used to tourists who used to stay for days or weeks in their hostels," said the owner of a hostel in São Jorge (INTERVIEW, FEB. 2017).

7. 1ENVIRONMENTAL ASPECTS CAUSED BY THE PAVING OF THE GO-239 MOTORWAY

In order to better analyse the changes that have taken place since the pavement was laid on the Estrada- Parque Prefeito Divaldo Rinco, on the stretch between São Jorge and Alto Paraíso de

Goiás, we captured the most relevant images and will discuss each one below.

This road can indeed be considered a Park Road, if it meets the characteristics required for such an object, as shown in Figure 1 of this study. However, it was pointed out, according to the president of the Alto Paraíso Green Integration Network, that "it doesn't have a management plan". However, Dourojeanni (2003, p. 78) points out that:

> the management of a Park Road is relatively simple and, although it is necessary to have a management plan, in general it does not require the detail and complexity typical of management plans for parks or other categories of direct or indirect use. [...] The management of a Park Road includes, on the one hand, the management of **the "road" and, on the other, the management of the "park".**

Figure 15, which shows the end of the asphalt on the GO-239, located just after the entrance to Vila de São Jorge, in the direction of Alto Paraíso de Goiás/São Jorge, shows an abrupt end to the asphalt and is located in a lowland, less than a kilometre from the Vila and close to the Park, from where you continue on to Colinas do Sul. There is no signpost at this location stating that this is the point where the tarmac ends. However, Installation Licence No. 1966/2013, granted to AGETOP by the Secretariat for the Environment and Water Resources, item 62, states that "in the area of direct influence of the PNCV, the maximum speed allowed on the road should not exceed 60 km/h, due to the time required for braking if there are animals on the road".

Figure 15: End of tarmac on GO-239 **Source:** The authors (2017)

What can also be observed is that, as it is downhill, there is a lot of stone and rock run-off, typical of the region, due to the fact that the tarmac ends in an area of rocky walls on the left-hand side, in the direction of Colinas do Sul.

Another observation is that there are no speed reducers or signposts in the area, so anyone coming from Alto Paraíso de Goiás is suddenly travelling on the dirt road. Given the speed we saw, over 80 km/h, there is an imminent danger of slipping, caused by the road ahead, as there is no shoulder and it is bumpy.

For Dourojeanni (2003, p. 77),

> Park Roads have, as an indispensable complementary element, a series of car parks and/or viewpoints strategically located at points of relevant landscape and/or wildlife interest. [...] A Park Road is distinguished by its signposting, which is not only typical of any road, but also has signposting and information about the natural features to be observed, the fauna, etc. Depending on the case, the speed may be normal for the characteristics of the road, or it may be much lower when the attractions are very important [...].

The sign in Figure 16 is located less than 500 metres from the entrance to São Jorge, in the direction of Alto Paraíso de Goiás. In a precarious state, rusty and graffitied, the writing is almost illegible and the wood that supports it is rotten.

Figure 16: Signpost
Figure 17: Tourist sign
Source: The authors (2017)

According to a local environmentalist, "the authorities responsible for preserving the plaques on the GO-239 don't give a damn about the state they're in. We, who love this place, are the ones who go around fixing these stakes" (INTERVIEW, FEB. 2017).

On the plaque in Figure 17, there are signs of gunshot wounds on the metal plate (on the right-hand side, below), and the wood is also rotten. Despite what it represents, this sign needs support, as it is crooked and about to fall over. It is also poorly located, as it is on a bend.

Located at Km 79 of the GO-239, a galvanised steel guardrail (Figure 18) was put in place to provide protection and prevent rollovers along the long curve of almost 200 metres, given that the terrain is a ravine.

Figure 18: Galvanised steel guardrail **Source**: The authors (2017)

Agetop's Licence, item 11, states that "metal fenders must be installed in the vicinity of the body of water, in order to prevent vehicles from falling into these bodies of water in both directions of the Special Works of Art or Current Works of Art".

According to a study by Dourojeanni (2003, p. 78),

With regard to the road, in addition to the normal precautions for the type of road, some special precautions must be taken. These include careful maintenance of signs and speed reduction and control equipment, elimination of roadside vegetation that reduces the visibility of the landscape, the constant presence of police patrols to prevent speeding, risky behaviour for visitors or wildlife, among others.

On this road, we only found two rubbish bins (Figure 19), both full, and the amount of rubbish scattered on the tarmac clearly showed that it hadn't been collected for days. We also noticed a certain number of soft drink and beer cans, as well as pet bottles and glass bottles, scattered along the road.

Figure 19: Rubbish on the roadside
Figure 20: Concrete manhole
Source: The authors (2017)

Figure 20 shows one of the three culverts for water drainage, which could also be used for the passage of wild animals. According to the AGETOP licence, "in drainage systems that are characterised by wildlife corridors, culverts with a minimum diameter of 2.00 metres should be planned", but these culverts do not reach this height.

7.1.1 Signalling on GO-239

Along the entire route, we noticed that every five kilometres there is a sign indicating that the route is a wildlife corridor and, depending on the corridor, there is a specification of the animal that passes through the place, as can be seen in Figure 21. This first sign, for example, warns that the predominant animal in the area is the Cachorro do Mato, while the others refer to the viado campeiro, the emu and the maned wolf.

There is also a sign every five kilometres indicating that speed should be reduced to 40 km/h, see Figure 22, as there are speed reducers nearby.

Figure 21: Fauna corridor
Figure 22: Speed reduction
Source: The authors (2017)

In item 61 of AGETOP's Technical Requirements, the following stands out:

> At the points selected by the environmental impact study as wildlife corridors, there should be speed reducers and a maximum speed of 40 km/h, as well as sound bells on both sides, which should be detailed in the Basic Environmental Plan (PBA)[6] . The signposting of the existence of the National Park and information on animal traffic should be more frequent.

Along this entire road, on the initiative of AGETOP ánd Goiás Turismo (GOIASTUR), eight speed reducers made of asphalt material (Figure 23) have recently been placed, in the form of low spring breakers, which are about ten centimetres high. Of these, two are near the bridges over the Rio das Cobras and Rio do Couro. "Although this has been a long-standing demand of the Vila's residents, it only happened now because during the celebrations at the end of 2016, a jaguar, which is endangered, was killed by being run over on this tarmac, near Alto Paraíso," says the head of the PNCV (Interview, FEB. 2017).

Figure 23: Speed reducer **Source**: The authors (2017)

The problem with these reducers, however, is that they were built according to the width of each lane, not covering the length of both sides and with a gap of about 5 metres between them. As we have seen

6 In the Installation Licence phase, the Basic Environmental Plan or Basic Environmental Project (**PBA**) contains details of the mitigating and compensatory measures to be adopted by the entrepreneur to mitigate environmental impacts identified in the EIA/RIMA.

on the ground, this causes motorists to drive in a zigzag pattern in order to get round them.

In high season or on long holidays, this is very likely to be "a major cause of accidents on this road, given that the idea is not seen as one of the best," said a resident of the village (Interview, FEB. 2017).

The sign in Figure 24, towards Alto Paraíso de Goiás/São Jorge, shows how vandalism is prevalent in the region in terms of lack of conservation and disrespect for road signs, as it is graffitied. The sign in Figure 25, which is about 20 kilometres away on the GO-239, still in the São Jorge/Alto Paraíso direction, also demonstrates the same circumstance. Figure 26 is of a sign installed near Jardim de Maytrea. It shows the illegibility of the lettering and the poor conservation of the sign.

Figure 24: Graffiti plaques
Figure 25: Graffiti plaque
Figure 26: Illegible sign
Source: The authors (2017)

7.1.   2Tourist attractions   located on the GO-239

The Pousada Fazenda São Bento, as shown in Figure 27, is eight kilometres from Alto Paraíso, is also entered via the GO-239 motorway and attracts a large number of visitors. Its attractions include waterfalls, zip lines and bike trails.

Figure 27: São Bento Farm
Figure 28: Valley of the Moon
Figure 29: Rancho do Waldomiro
Source: The authors (2017)

At a distance of 9 kilometres from São Jorge, by asphalt (5 km) and dirt road (4 km), and by trail of about two kilometres, Vale da Lua (Figure 28) is one of the most visited attractions in the PNCV. This place is known for its rock formations, which are carved by strong rapids that form craters reminiscent of a lunar landscape.

Rancho do Waldomiro (Figure 29) is one of the most traditional places in the region, because as well as being on the main road linking Alto Paraíso to the PNCV, ten kilometres from São Jorge, the place is famous for offering tropeiros' food, better known as Matula. This dish is prepared with common beans in the form of a tutu seasoned with saffron and four other types of tinned meat. It is served in a banana leaf, plus the following side dishes: paçoca de carne seca (farofa), fried manioc, boiled pumpkin, rice and tomatoes.

Figure 30: Maytrea Garden
Figure 31: Morro da Baleia
Figure 32: Buracão Hill
Source: The authors

According to Dourojeanni (2003, p. 77), "the infrastructure for visitors to the Park Roads consists of viewpoints located in the car parks themselves, from where, if necessary, footbridges or trails can lead to places of interest that are further away".

As stated in item 59, also in Process 19224/2011 (AGETOP), "the motorway may not have 'stopping points' for tourists, with the exception of the existing 'Maytrea Garden' (Figure 30), which should only receive treatment to contain erosion processes". However, this site has a rather precarious infrastructure, with no space for parking or even a deck to enjoy the garden.

According to research carried out on social media (http://www.eco.tur.br, 2017), the first of the attractions, the Morro da Baleia viewpoint (Km 20, Figure 31), was once known as Morro do Ferro de Engomar. At an altitude of 1,500 metres, it is one of the PNCV's postcards. It doesn't have a wooden deck, but there is a three-kilometre trail and from the top you can see part of the city of Alto Paraíso de Goiás, as well as several mountain ranges that border the site.

Below, at the foot of the Morro da Baleia mountain range, we have what is known as the Maytrea Garden (Km 20), which is actually several buritis trees forming what is known as a vereda. This landscape demonstrates the relief of the plateau.

A humid field with springs that form the Riacho Fundo stream, a tributary of the Rio Preto, which cuts through the entire region, this garden is a stop for photos and landscape appreciation for most tourists heading towards São Jorge.

The third point, the Morro Buracão viewpoint (Figure 32), located at Km 21 of the road, is a difficult place to access, with steep trails and a quartzite rock formation. Despite this, the attraction does not go unnoticed by backpackers who frequent the region or even more adventurous tourists.

Figure 33: Dead bird on GO-239
Source: Fragoso (2017)

It's not uncommon to find dead and run over birds (Figure 34) on the track of the Divaldo Rinco Park Road. When we were in the area, it rained for three consecutive days. Due to this factor, we believe that the water carried away the animals that were run over during this period, but it is always possible to find crushed animals along the route.

CHAPTER 8

Between 11/02/17 and 15/02/17 we went to the district of São Jorge to interview some of the administrators, residents, tourists, traders and guides who live in and visit the area, as well as to observe and photograph the Divaldo Rinco Park Road, as shown above.

According to the Environmental Analyst for the Ministry of the Environment and head of the PNCV, the paving of the GO-239 took place through a process that had already been established. "I lived in São Jorge for the first time in 2003 and there was no tarmac, but I already had the project and the news that it would come. It was determined by the state government." (Interview, 2017)

When asked how this road is used, he says that "for traffic from São Jorge to Alto Paraíso de Goiás and vice versa, by residents of Colinas do Sul and also by tourists from the PNCV and fishing tourists from Serra da Mesa[7] ".

We asked Mr Fernando what, in his opinion, makes the GO-239 a

Park Road:

> The decree that gives it its name. EP is not a category of Conservation Unit in the SNUC or the SEUC (State System of Conservation Units). There is no such category. So, as far as I'm concerned, it's just a name. For it to deserve this name, what leads it to have this name, is that it tangents and cuts through the PNCV at seven points. What gives it this name is the scenic beauty around it, the landscapes of the Chapada. (Interview, FEB. 2017)

The next question involved the topic of what has been done to turn the GO-239 into a Parkway. According to the Environmental Analyst,

> There was a movement, a mobilisation of civil society to reduce the very high number of people being run over by wildlife, all of this prior to the implementation of the speed bumps, and in my opinion, this is what led to their implementation. There's an NGO called Associação Amigos das Florestas (Friends of the Forests Association) that met with the president of Agetop, with João Lino, GOIASTUR's project manager, at the end of 2016 and submitted a letter calling for the inclusion of measures to monitor the trampling of fauna in the PAN - National Action Plan for the Conservation of Endangered Wild Canids - (an acronym for various groups of endangered species) and the PNCV is participating in this plan. There are 4 species of wild canids[8] threatened with extinction in Brazil, three of which occur here and there are records of them being run over. Of these species here: maned wolf, field fox and vinegar dog. So monitoring measures are provided for in this PAN and management measures with the motorway's managing body, AGETOP. Monitoring hasn't been designed yet, but we're thinking of using PNCV staff. For example, there are five park employees who live in Alto Paraíso, so they travel this stretch almost every day. So they would make this observation, which would be to record which species has *been* run over and where." (Interview, FEB. 2017)

---

[7] **Serra da Mesa Lake** (the artificial lake of the **Serra da Mesa Power Plant**) is the fifth largest lake in Brazil, located in the north-west of Goiás. It is about 95 kilometres from São Jorge towards Colinas do Sul.

[8] **Canids:** Canids are a family of digigrade mammals, of the Order of Carnivores, which includes the Dog, Wolf, Coyote, Jackal, Mabeco, and Fox among others. This family is sometimes divided into two tribes: Canini and Vulpini.

According to the head of the PNCV, the decision to place the reducers came about because "there are nine points, including one that is wrong. Even without systematic monitoring, we see a higher rate of accidents at these points". He added: "It was an outsourced company that built them, but AGETOP ordered it. Any intervention has to be done by them."

To finalise the interview, Fernando Tatagiba explained his thoughts on the project to turn the GO-239 into a Parkway:

> I think it can work, but ideally, before implementing the measures, there should also be a discussion with the residents or with the councils. There are public participation forums here, for example: we have a tourism secretary and an environment secretary. There is a council for each of these areas. There's the Pouso Alto APA council, which is a State Conservation Unit. There's a Park Council. As the road cuts through the Park in seven sections, the ideal thing would be for the interventions on the road to be brought before the Park's council. This is to improve the construction of this project, so that the EP is more than just a name. There is no concept of EP established in any legislation [...]. In order to deserve this name, the ideal is for there to be some criteria for declaring it an EP and for impact measures to be put in place, in short, to take into account both environmental and tourism issues. João Lino has already spoken to us about the idea of building viewpoints on the road. (Interview, FEB. 2017)

Another participant in our research was the Secretary of Tourism of Alto Paraíso de Goiás. For her, the change from GO-239 to Estrada-Parque is positive because:

> The practice of sports, such as cycling, on that stretch has increased mainly among tourists from Brasilia. But there are many families from the region practising the sport too, once the road has been improved. In addition, there would be a drastic reduction in animals being run over on the road." (Interview, FEB. 2017)

A resident of Vila de São Jorge for over ten years, journalist Zanoni Antunes was one of the people interviewed to answer questions about the district's residents.

Asked which tourist attractions are considered the most attractive on the road from São Jorge to Alto Paraíso, he replied that "without a doubt they are Vale da Lua and Fazenda de São Bento".

The next question was how he saw the landscape on this route, to which he replied:

> There is rampant urban sprawl that urgently needs to be controlled. Agribusiness and the lack of rules on deforestation are destroying this beautiful roadside landscape. Those who work in agriculture here don't respect the environment, because they even plant on hills and in sloping areas, causing pesticides to run into rivers. (Interview, FEB. 2017)

Throughout our field research, we observed that most of the interviewees were not even aware that the GO-239 was a Parkway or what that might mean. We also observed that they were completely unaware that the road had a name and what that name was, or even who Sulivan Silvestre was, the first name given.

In our research with some of the village's residents and guides, those who are aware of what a park road is say that this is only because it has a scenic beauty that differs from other roads, but they can't translate what this beauty is like.

When asked what can be seen about the road on social media, few said they had seen anything about it on the Internet, adding that it only appears on television when animals are run over or accidents occur, such as the death of a tour guide at the end of 2016.

João Trindade, an anthropologist who has lived in São Jorge for more than five years, was one of the tourist drivers who gave us some feedback. For him, the impression you get when you drive along the stretch between São Jorge and Alto Paraíso de Goiás "is the quality of the tarmac, which is like a carpet. The cycle path is beautiful, and it illuminates the road at night". For the guide, there is a lack of signposting at the end of the road, where the tarmac ends and the stretch linking São Jorge to Colinas do Sul begins. "This is the cause of many skids. There have even been rollovers there," he said, confirming what he had observed.

CHAPTER 9

FINAL CONSIDERATIONS

The GO-239 Estrada-Parque Prefeito Divaldo Rinco is the road that gives access to the PNCV, located between the municipality of Alto Paraíso de Goiás and the district of São Jorge.

This road has unrivalled scenic beauty. Although this road runs as far as the town of Colinas do Sul, 86.5 kilometres from Alto Paraíso, only the 36 kilometres that are paved were analysed in this study.

The tarmac is in good condition. The average speed allowed on the road is 80 kilometres per hour, but this is reduced to 40 kilometres per hour where there are wildlife corridors, and 60 kilometres per hour when travelling in a straight line. However, due to the lack of enforcement, the speed limit is rarely respected.

On this route, we can see that there are eight wildlife corridors, as the signs show. At these points, speed reducers have been built with asphalt material in the shape of small undulations on just one half of the road.

This motorway does not have any tunnels for animals to pass through, but there are culverts for water run-off that can be used by small animals when crossing the road.

On the road there is a cycle path and two viewpoints, which are signposted but do not have a place to turn around and park your car to facilitate tourist activities.

Tourist attractions such as Vale Da Lua, Fazenda São Bento and Rancho do Waldomiro also have signs signposting them.

As we can conclude, the population of São Jorge sees the road in question only as the road that leads to Alto Paraíso. When asked what a Park Road is, most of them don't even know what it means. From the point of view of the PNCV administrator and a journalist who was also interviewed, this road has the legal backing of a Park Road, but theoretically it doesn't work as described in its characteristics, i.e. it is characterised as such, but it doesn't yet have a management plan.

This road has great tourist potential given the scenic beauty that surrounds it. It is a Parkway that exists in law, but not in fact.

We believe that if administrators, environmentalists and even the community got together to draw up a management plan that favoured this stretch of road, highlighting its real value, it would become an additional attraction for the region.

REFERENCES

BARRETO, Margarita. Manual de Iniciação ao estudo do turismo. 13ª Ed. Campinas, SP. Papirus, 2003. Tourism Collection. 160p.

BEHR, Miguel von. Cradle of waters and the new millennium: Chapada dos Veadeiros, Goiás, Brazil. Brasília: Editora UnB: Editora IBAMA, 2000.

BOLSON, Jaisa H. Gontijo. The importance of landscape in tourism.
Available at http://www.revistaturismo.com.br/artigos/paisagem.html. Accessed on 20/07/2016

BARROS, Lídia Almeida. Vocabulário enciclopédico das Unidades de Conservação do Brasil. São Paulo,
SP: Ed. Unimar; Arte e Ciência. 2000.

CRUZ, R. C. Política de Turismo e Território. São Paulo. Contexto, 2000.

CONFLORA. Environmental Impact Report for the asphalt paving of the GO-239 motorway. Goiás:
Conflora, 2004.

DIAS, Reinaldo. Sustainable tourism and the environment. São Paulo: Atlas, 2003. 208p.

Introduction to tourism. São Paulo: Atlas, 2005. 178p.

DIAS, C. M. M. "Home *OEvy from* Ãome": Evolution, characterisation and perspectives of the hotel
industry - a comprehensive study. Master's dissertation. Eca.-USP. São Paulo, 1990.

DOMICIANO, Carlos Shiley. Environmental values and development: a case study of the District of São
Jorge and the Chapada dos Veadeiros National Park. 2014. 203f. Thesis (Doctorate in Environmental
Sciences). Federal University of Goiás, Postgraduate Programme in Environmental Sciences.

DOMICIANO, Carlos Shiley; OLIVEIRA, Ivanilton José. Mapping environmental impacts in the Chapada
dos Veadeiros National Park (GO). Fortaleza: Mercator UFC. 2012.

DOUROJEANNI, M. J. Park roads, an under-explored opportunity for tourism in Brazil. Nature &
Conservation. 2003. 74-77

EMBRATUR/IBAMA. Guidelines for a national ecotourism policy. Brasilia: MICT/MMA inter-ministerial
working group, 1994.

ENVIRONMENTAL IMPACT STUDY - EIA/RIMA. GO-239 motorway, sections between GO 118/BR-
010 and GO-132. Goiás State Government, Department of Transport and Public Works. Department of
Highways. CONFLORA, 1998.

S.O.S ATLANTIC FOREST FOUNDATION. Park Road: concept, experiences and contributions. São
Paulo, 2004. 60p.

FERREIRA, Mariana V. N. R. **Tourism and air transport policies in Brazil since the 1990s.** FACCT. Niterói,
2009.

GOMES, Edvania: **Landscape, Imaginary and Space**. Rio de Janeiro: Ed. UERJ, 2001.

HAUFF, Shirley Noely. **Relations between local rural communities and park administrations in Brazil**:
subsidies for the establishment of buffer zones. Thesis presented as a partial requirement for the degree of
Doctor. Postgraduate programme in Forestry Engineering, Nature Conservation area, Agrarian Sciences
sector, Federal University of Paraná. 225p. Curitiba, 2004.

BRAZILIAN INSTITUTE FOR FORESTRY DEVELOPMENT - IBDF / BRAZILIAN FOUNDATION
FOR NATURE CONSERVATION - FBCN. **Plan for the Brazilian Conservation Units System**: Stage II.
Brasilia, 1982.
IBGE, Brazilian Institute of Geography and Statistics. **Demographic census: 2010**. Rio de Janeiro, 2010.

LACERDA, Roberta Carolina Lima Gontijo. **The impacts of tourism on the perception of the community of Vila de São Jorge**: the gateway to the Chapada dos Veadeiros National Park, Goiás. 124p. 11/2007. Master's dissertation - UNA University Centre, Master's Programme in Tourism and the Environment. Belo Horizonte, 2007

MAGALHÃES, Cláudia Freitas. **Guidelines for sustainable tourism in municipalities.** São Paulo: Roca, 2002.

MAIA, Yuri de Carvalho. **Evaluation of the impacts of paving the Estrada-Parque in the villages of Visconde de Mauá, Rj.** Course Conclusion Programme. Fluminense Federal University, RJ. 2014.
Available at: <http://www.repositorio.uff.br/jspui/bitstream/1/1058/1/294%20- %20Yuri%20Maia.pdf>.
Accessed on 20/10/2016.

MENDONÇA, Dálio Filho. **The study of ecotourism in the Chapada dos Veadeiros in the state of Goiás, Brazil.** A strategic environmental vision. Brasília, 02/2007. 111p.
Master's thesis. Centre for Sustainable Development, UNB, Brasília, 2007. MMA - Ministry of the Environment. **Strategic environmental assessment.** Brasília, 2002.

NATIONAL SCENIC BYWAYS PROGRAMME. List of activities. Available at : <http:// www.byways.org/explore/activities/> Accessed on 20/07/2016.

OLIVEIRA, Antônio Pereira. **Tourism and development**: planning and organisation. 4ª Ed. São Paulo: Atlas: 2002. 287p.

WORLD TOURISM ORGANISATION (Madrid and New York). International Recommendations for Tourism Statistics. Journal of the World Tourism Organisation, Madrid and New York. P. 1-31, 15 Jan. 2009

OXFORD UNIVERSITY PRESS. The New Shorter Oxford English Dictionary. Sixth Edition, 2007. 3,472 pages.

PIRES, Paulo dos Santos; TIAGOR, Aline Almeida de. **The tourist potential of park roads at the confluence of landscape and protected areas.** Available at:
<www. anptur.net> . Accessed on 20/07/2016.

PARKWAY PARTNER ORGANISATIONS. History of the Parkway. Available at:
<http://www.blueridgeparkway75.org/> Accessed on 18/08/2016

CHERUBIM MAGAZINE. **Tradition and sustainability:** a study of traditional cerrado knowledge in Chapada dos Veadeiros, Vila de São Jorge - GO. Issue 09, no. 21, Vol 02. Rio de Janeiro, 160 p.

BRAZILIAN JOURNAL OF TOURISM RESEARCH. **Park roads:** a comparative study towards definitions for the Brazilian tourist experience. São Paulo. Issue Jan/April 2012. Page 79-94

REJOWSKI, M. **Turismo e pesquisa científica:** pensamento internacional x situação brasileira. Campinas, Sp. Papirus, 1996.

RODRIGUES, Adyr Balastreri. **Turismo e espaço: rumo a um conhecimento transdisciplinar**, 2. Ed. São Paulo: Hucitec, 1999.

RUSCHMANN, Doris Van de Neebe. **Tourism and sustainable planning:** Protecting the environment. Campinas, SP: Papirus, 1997. (Tourism Collection)

SANTOS FILHO, J. **Mirror of History:** the phenomenon of tourism in the course of humanity. Espaço Acadêmico Magazine. Maringá/PR, V, 50. Available at <www.espacoacademico.com.br> Accessed on 04/01/2017

SANTOS, Niara; MARTINS, Alessandra. **Basic information on Protected Areas Environmental - APAs and Park Roads.** Available at:
<http://www.amigosdemaua.net/estrada/documentos/APAs%20e%20Estradas- park.htm.>Accessed 18/08/2016

SILVA, Clarinda Aparecida da. **Landscape - Field of visibility and socio-cultural significance:** Chapada dos Veadeiros National Park and São Jorge Village. 2003. 186 f. Thesis (Master's Degree in Geography). Federal University of Goiás, Institute of Socio-Environmental Studies.

SILVA, Elisangela A. Machado. **Socio-economic effects of transport infrastructure in tourist locations:** Paving the GO-239 in Vila de São Jorge. USP, 2005.

SILVA, Lauro Leal. **Ecology:** wilderness management. Santa Maria, RS: Ministry of the Environment; Science and Technology Support Foundation (FATEC), 1996.

SORIANO, Afrânio José Soares. **Park Road:** a proposal for a definition. 20/11/2006. 181p. Thesis (Doctorate) - São Paulo State University, Institute of Geosciences and Exact Sciences.

SOLHA, K. T. Evolução do Turismo no Brasil, in: Rejowski, M. (org.) **Turismo no percurso do tempo.** São Paulo. Aleph, 2002.

STRAUSS, A.; Corbin, J. Qualitative research: techniques and procedures for developing grounded theory. 2ª ed. Porto Alegre, Artmed, 2008.

TASCHNER, Gisela B. Leisure and tourism in Brazil: notes on its trajectory. FGV, 2003.

TRICÁRIO, L. T., OLIVEIRA, J. P., Rossini, D. M., Carvalho, D. I. Revista Brasileira de Pesquisa em turismo: Estradas-parque: A comparative study towards definitions for the Brazilian tourist experience. São Paulo, 6th Ed. PP. 79-94. Jan/Apr. 2012.

WEARING, Stephen; NEIL, John. Ecotourism: impacts, potential and possibilities. Barueri: Manoel, 2001.

WEBSITES ACCESSED:

The romantic route. Available at:
<http://magazin.outdooractive.com/de/2010/03/22/die-romantische-strase-wird-60/. Santos, M. (1996) > Accessed on 18/08/2016.

Tourism Brazil - Reference 2011-2014. Available at:
< http://www.portalbrasil.net/downloads/TurismoBrasil_Referencial_2011_2014_Mtur.pdf> Accessed on:11/01/16

Aurélio Basic Dictionary of the Portuguese Language. Available at:
<https://dicionariodoaurelio.com> Accessed May/2016

ECO.TUR.BR, 2012. Vila de São Jorge. Available at:
<http://www.eco.tur.br/ecoguias/veadeiros/roteiros/sjorge.htm> Accessed on 12/08/2016.

Sports and Leisure, 2010. Available at:
<http://www.itacare.com/itacare/megabusca.php?group=ecotrip&cat=trilha +cachoeira&lang=portugues> Accessed on 12/08/2016.

Pantanal Park Road. Available at:
<http://www.webventure.com.br/destinoaventura/destinos/index/atracoes/destino/corumba/atr /395>
Accessed on 18/08/2016.

Pantanal Park Road. Available at:
<http://blogandoturismo.googlepages.com/CAPTULOII.PDF> Accessed on: 16/08/2016.

Germany's Romantic Road.
Accessed at: <www.rodgerrealm.com/germany> Accessed on 16/08/2016.

GOIÁS. State Secretariat for Planning Development Superintendence of Statistics, Research and
Information. State Law No. 4685 of 1963. Available at: http://cidades.ibge.gov.br/painel/historico.
php?codmun=520060>. Accessed on 19/12/16

Michaelis English School Dictionary. Available at: <http://michaelis.uol.com.br/moderno-ingles/> Accessed
May/2016

Cantareira State Park.
Available at: <http://turmadoarboreto.blogspot.com/2007/12/parque-estadual-da- cantareiraa-maior.html>
Accessed on 16/08/2016.

Serra da Cantareira State Park.
Available at:
http://www.guiadasemana.com.br/Sao_Paulo/Passeios/Estabelecimento/Parque_Estadual_da_
Serra_da_Cantareira.aspx?id=1151 Accessed on 16/08/2016.

Tourist attractions in the region.
Available at: <http://www.elkaris.com/?pg=14&keys=Itacare+Pontos+
Turisticos&t=1258> Accessed on 16/08/2016.

Portal of Alto Paraíso, Goiás.
Available at: <http://www.altoparaiso.go.gov.br/Geografia.php> Accessed on 18/08/2016.

_____________________ Available at:

<http://www.altoparaiso.go.gov.br/Historia.php> Accessed on 16/12/2016.

ICMBio Portal. Available at:
<http://www.icmbio.gov.br/parnachapadadosveadeiros/guia-do-visitante.html> Accessed on 05/01/16

Hotel magazine. Available at:
<http://www.revistahoteis.com.br/a-decada-de-transicao-da-hotelaria-e-do-turismo- brazilian/> Accessed on
11/01/16

Romantische Strasse . Available at:
< http://romantischestrasse.de/. > Accessed on 18/08/2016.

SNUC - National System of Conservation Units. Federal Law No. 9985/2000.
Available at:
<http://www.planalto.gov.br/ccivil_03/leis/L9985.htm.> Accessed 10/11/16

APPENDIX A - Interview questionnaires

- Residents

1.   In your opinion, which attractions draw tourists' attention on the Alto Paraíso/São Jorge/PNCV route (apart from the viewpoint)?

2.   How do you think the following items work on the São Jorge / Alto Paraíso route?

a.   Traffic (are there restaurants, cycle lanes or cycle paths? Is there a continuation or exit to somewhere?) Is there a pedestrian walkway or safety lanes?

b.   What is the most common means of transport (cars, buses, minibuses, carts, etc.) to get to Vila de São Jorge?

d. What is your impression of the road from Alto Paraíso to the PNCV? How do you see the landscape along the way? Is there human interference? Is the scenery predominantly natural?

3.   About signage and visual communication (do the signs match the adverts? What do you notice?)

4.   What are the fauna (animals) and flora (plants) like on this route?

5.   What do you see on the internet about this road?

6.   What do the residents of São Jorge or the surrounding area call the road?

7.   Can you tell us why this road was called the Park Road?

- Administrators

1.   Can you tell us how the asphalting of the GO-239 (Estrada Parque Prefeito Divaldo Rinco) took place, when it was completed and what the purpose of this asphalting was?

2.   How is the road between Alto Paraíso and São Jorge used (for leisure, tourism, attractions, freight transport, etc.)?

3.   In your opinion, what makes the GO-239 a Park Road (the fauna and flora)? Is tourism publicised along this route? In what way?

4.   What do you see on the internet about this road? Have you read anything about it?

5.   What do the residents of Alto Paraíso/São Jorge or the surrounding area call the GO-239?

6.   What has been done in favour of transforming the GO-239 into a Parkway, a since by law it is already labelled as such?

7.   What do you think about this project in particular?

8.   Can you tell us if there are any plans to create tolls on this motorway?

ANNEX A - Decree-law for the first name of the road

LEI Nº 13.467, DE 20 DE JULHO DE 1999.

Denomina o trecho rodoviário que especifica.

A ASSEMBLÉIA LEGISLATIVA DO ESTADO DE GOIÁS decreta e eu sanciono a seguinte lei:

Art. 1º Fica denominada ESTRADA PARQUE PREFEITO DIVALDO RINCO a Rodovia GO-239, no trecho que liga o Município de Alto Paraíso ao Município de Colinas do Sul.
• Redação dada pela Lei nº 19.015, de 22-09-2015.

~~Art. 1º - Denomina-se ESTRADA PARQUE SULIVAN SILVESTRE o trecho da GO-239, que passa em frente ao Parque Nacional da Chapada dos Veadeiros, localizado entre os Municípios goianos de Colinas do Sul e Alto Paraíso.~~

Art. 2º - Esta lei entrará em vigor na data de sua publicação, revogadas as disposições em contrário.

PALÁCIO DO GOVERNO DO ESTADO DE GOIÁS, em Goiânia, 20 de julho de 1999, 111º da República.

MARCONI FERREIRA PERILLO JÚNIOR
Floriano Gomes da Silva Filho
Sebastião Monteiro Guimarães Filho

(D.O. de 30-07-1999)

Este texto não substitui o publicado no D.O. de 30.07.1999.

ANNEX B - Decree-law for the second name of the road

GOVERNO DO ESTADO DE GOIÁS
Secretaria de Estado da Casa Civil

LEI Nº 19.015, DE 22 DE SETEMBRO DE 2015

Altera a Lei nº 13.467, de 20 de julho de 1999, que denomina o trecho rodoviário que especifica.

A ASSEMBLEIA LEGISLATIVA DO ESTADO DE GOIÁS, nos termos do art. 10 da Constituição Estadual, decreta e eu sanciono a seguinte Lei:

Art. 1º O art. 1º da Lei nº 13.467, de 20 de julho de 1999, passa a vigorar com a seguinte redação:

"Art. 1º Fica denominada ESTRADA PARQUE PREFEITO DIVALDO RINCO a Rodovia GO-239, no trecho que liga o Município de Alto Paraíso ao Município de Colinas do Sul." (NR)

Art. 2º Esta Lei entra em vigor na data de sua publicação.

PALÁCIO DO GOVERNO DO ESTADO DE GOIÁS, em Goiânia, 22 de setembro de 2015, 127º da República.

MARCONI FERREIRA PERILLO JÚNIOR
Vilmar da Silva Rocha

(D.O. de 28-09-2015)

Este texto não substitui o publicado no D.O. de 28-09-2015

ESTADO DE GOIÁS
SECRETARIA DO MEIO AMBIENTE E DOS RECURSOS HÍDRICOS

Licença de Instalação

Processo: 19224/2011	**Licença: 1966/2013**

A SECRETARIA DO MEIO AMBIENTE E DOS RECURSOS HÍDRICOS DO ESTADO DE GOIÁS, no uso de suas atribuições que lhe foram conferidas pela Lei Estadual n.º 8.544, de 17 de outubro de 1978, regulamentada pelo Decreto 1.745/79, concede a presente LICENÇA DE INSTALAÇÃO, nas condições especificadas abaixo:

Cliente

1. Razão Social: **AGENCIA GOIANA DE TRANSPORTES E OBRAS**
2. CPF/CNPJ: **03.520.933/0001-06**
3. Endereço: **AVENIDA GOVERNADOR JOSE LUDOVICO DE ALMEIDA, nr. 20, .CAIÇARA**
4. Município: **Goiânia - GO**

Empreendimento

1. Razão Social: **PAVIMENTAÇÃO DA RODOVIA GO 239**
2. CPF/CNPJ:
3. Endereço: **RODOVIA GO 239, nr. S/N, S, ZONA RURAL**
4. Município: **Alto Paraíso de Goiás - GO**

Bacia Hidrográfica/ Micro Região

1. Bacia Hidrográfica: **Tocantins**
2. Micro Região: **Chapada dos Veadeiros**

Atividade Licenciada

1. Nome: **PAVIMENTAÇÃO DE ESTRADA**

Parâmetros

1. Extensão: **70,00km**

Exigências Técnicas - Observações

1. A presente Licença está sendo concedida com base nas informações constantes do processo e não dispensa e nem substitui, outros alvarás ou certidões exigidas pela Legislação Federal, Estadual ou Municipal;

2. A SEMARH deverá ser comunicada, imediatamente, em caso de acidentes que envolvam o Meio Ambiente;

3. A SEMARH reserva-se o direito de revogar a presente Licença no caso de descumprimento de suas condicionantes ou de qualquer dispositivo que fira a Legislação Ambiental vigente, assim como, a omissão ou falsa descrição de informações relevantes que subsidiam a sua expedição, ou superveniência de graves riscos ambientais e de saúde.

4. Fica a presente automaticamente SUSPENSA, independente de qualquer ato administrativo por parte desta Secretaria, caso expire o prazo de validade das demais licenças emitidas por outros entes da Administração Pública, seja municipal, estadual ou federal, que fazem parte da instrução do processo a que esta se vincula. Somente com a juntada nos autos de novo documento que será restaurada a validade da licença ora emitida;

5. Deverão ser preservadas as faixas previstas na Lei n.º 12.596/95 como Áreas de Preservação Permanente, sendo inclusive vedado qualquer tipo de impermeabilização do solo;

6. A Licença de Funcionamento deverá ser requerida 30 (trinta) dias antes do início previsto para operação, ficando sua concessão condicionada às exigências técnicas constantes do verso desta Licença;

7. Esta licença não produz efeitos jurídicos de cessão e/ou aquisição sobre direito de posse e direitos reais como: de propriedade (uso, gozo e disposição), de superfície, de usufruto, de servidão, de habitação, de uso, de penhor, de hipoteca, de anticrese e direito do promitente comprador de imóvel; bem como demais direito inerentes à propriedade móvel e imóvel sobre a área e bens delimitados e discriminados nesta licença; nem mesmo direito adquirido, produzindo somente efeitos jurídicos nos limites da Legislação Ambiental e de competência da SEMARH dentro de seu poder de polícia preventivo e repressivo.

8. A renovação da presente Licença deverá ser requerida com antecedência mínima de 120 (cento e vinte) dias da expiração de seu prazo de validade, ficando este prorrogado até a manifestação definitiva deste órgão.

9. Conforme disposto na Resolução CONAMA 006/86, o licenciado deverá providenciar a publicação do recebimento da presente renovação de licença de instalação no prazo de 30 (trinta) dias a partir desta data, podendo a mesma ser suspensa, caso não haja cumprimento desta;

10. A AGETOP, deverá requerer em até 30 (trinta) dias após a conclusão dos serviços da obra a Licença de Funcionamento da Implantação e Pavimentação da Rodovia GO-239, neste trecho licenciado;

11. A AGETOP, deverá requerer em caso de não conclusão dos serviços da obra no prazo de vigência desta licença de instalação, a renovação da licença de instalação com antecedência mínima de 120 (cento e vinte) dias antes de expirar a sua vigência;

12. A paralisação temporária ou a conclusão das atividades do Implantação e Pavimentação da Rodovia GO-239, deverá ser objeto de comunicação a esta SEMARH.

Exigências Técnicas - Complementares

1. Esta Licença de Instalação, refere-se as Obras de Implantação, Restauração, Readequação, Obras de Arte Corrente, Obras de Artes Especiais, Pavimentação, Sinalização Vertical e Horizontal da Rodovia GO-239, no Trecho de Alto Paraíso até o Distrito de São Jorge e a cidade de Colinas do Sul, com extensão de 70 quilômetros. A Requerente não deverá Ultrapassar os limites da área licenciada;

2. Manter as obras de acordo com o previsto no cronograma físico, considerando como mês inicial das obras o mês ou os meses subsequentes do recebimento deste Licenciamento, pela AGETOP;

3. Executar os serviços em horários apropriados, de forma a não prejudicar os moradores da região da Implantação e Pavimentação da Rodovia GO-239;

4. Manter, durante a fase das obras e após o início de funcionamento da Implantação e Pavimentação da Rodovia GO-239, a emissão de material particulado, ruídos e vibrações dentro dos parâmetros da Legislação Ambiental;

5. Utilizar os materiais obtidos nas limpezas das faixas de domínio e do decapeamentos dos locais de implantação da rodovia, para retaludamento das caixas de empréstimos e nos enchimentos destas;

6. Proteger os solos expostos e ainda não protegidos por meio do plantio de gramíneas, com lonas plásticas para evitar processos erosivos e de lixiviação dos solos, carreamento destes para drenagem superficial e profunda e destas para os leitos dos corpos hídricos o que poderá causar assoreamentos;

7. A requerente, deverá manter os Taludes dos Cortes e Aterros suavizados, atendendo a declividade máxima de acordo com o tipo de solo local e realizar hidrossemeadura com gramíneas de baixo crescimento, ao longo de todos os taludes para se evitar Processos Erosivos, e ainda se evitar obstruções na sinalização vertical da Rodovia;

8. Promover o disciplinamento das águas pluviais e prover o sistema de drenagem pluvial superficial de dissipadores de energia no local de deságue, com vistas a evitar processos erosivos ao longo de todo trecho;

9. Implantar e manter os dissipadores de energia nos pontos de lançamento de águas pluviais definidos no projeto de drenagem pluvial;

10. Monitorar os pontos de lançamentos finais das águas pluviais captadas na área da Implantação e Pavimentação da Rodovia, para prevenção de formação de processos erosivos.

11. Instalar defensas metálicas nas proximidades dos corpos hídricos, com vistas a inibir queda de veículos, nestes corpos hídricos nos dois sentidos das Obras de Arte Especial ou Obra de Arte Corrente;

12. Deverão ser instaladas sinalização vertical, com placas específicas, visando a mitigação de danos que podem ser causados a fauna local;

13. Deverão ser instalados sinalizadores progressivos em ambas as faixas para alertar os motoristas e provocar efeito de velocidade crescente, em pontos específicos, em especial nos corpos hídricos, app's, reservas legais e matas nativas;

14. Orientar todos os colaboradores envolvidos nas Obras da Rodovia, quanto aos aspectos de preservação ambiental, no que diz respeito à destinação correta de resíduos sólidos gerados, manutenção da vegetação nativa e demais práticas que melhorem o ambiente de trabalho, a segurança ocupacional e o convívio com a vizinhança;

15. Providenciar EPI's - Equipamentos de Proteção Individual, para todos os colaboradores envolvidos nas Obras de Implantação e Pavimentação da Rodovia;

16. Implantar sistemas de controles de trânsito de veículos e pedestres, com objetivo de minimizar a ocorrências de acidentes na área de influência direta dos serviços de Implantação e Pavimentação da Rodovia;

17. Definir destinação adequada aos resíduos sólidos e líquidos, provenientes dos equipamentos utilizados nos serviços, tais como filtros de óleo, filtros de ar, mangueiras, pneus, sucata ferrosa, óleos lubrificantes e outros tipos de resíduos;

18. Providenciar destinação adequada ao local definido pelos municípios de Alto Paraíso e Colinas do Sul para descarte dos RSCC - Resíduos Sólidos da Construção Civil que serão gerados durante os serviços das Obras de Implantação e Pavimentação da Rodovia;

19. Não lançar no solo sob hipótese alguma sobras de concretos (de caminhões betoneira ou de betoneiras móveis) utilizados nas obras de artes correntes e especiais, em específico drenagens superficiais, bueiros, pontes e outras, ao longo do trecho que está sendo licenciado, sendo necessário dar destinação adequada a estas sobras;

20. Não lançar nos solos da faixa de domínio ou área não autorizada, sob hipótese alguma sobras de massa asfáltica, utilizadas nas obras de Pavimentação do trecho que está sendo licenciado, sendo necessário dar destinação adequada a estas sobras;

21. Destinar adequadamente os eventuais efluentes gerados nas frentes de serviços, procurando sempre que possível a utilização de Banheiros Químicos para serem utilizados pelos colaboradores;

22. Manter sempre em boas condições de uso e de funcionamento os banheiros químicos, manter firma para limpeza e destinação adequada dos dejetos produzidos durante os serviços das Obras de Implantação e Pavimentação da Rodovia;

23. Os tanques com materiais betuminosos e óleos deverão ser dotados de sistema de segurança e proteção contra vazamentos, para que não ocorra a contaminação dos solos e das águas;

24. Que o local destinado para manutenção dos equipamentos, troca de óleo e armazenamento de combustíveis seja adequado não permitindo a contaminação do solo e das águas;

25. Realizar o armazenamento adequado de todos os óleos lubrificantes usados ou contaminados, gerados pelos equipamentos em utilização na implantação das obras, e posteriormente deverá ser enviado para firma de rerrefino, devidamente licenciada pelo órgão ambiental, dé acordo com o estabelecido no Art. 3 da Resolução CONAMA nº 362/2005.

26. Atender as normatizações da ABNT, quando for realizar as escavações, procurando sempre mitigar os fatores que possam gerar processos erosivos na Área de Influência Direta, os reaterros deverão esta bem compactados e o solo protegido contra processos erosivos nas proximidades dos pontos onde forem executadas as escavações;

27. Manter, durante e posteriormente aos serviços das obras, os cuidados e medidas de conservação dos solos a fim de evitar, formações de processos erosivos e de lixiviação dos solos, não o impermeabilizando, e promovendo sua proteção superficial, com a plantação de gramíneas de baixo crescimento ou outro procedimento adequado para as áreas a serem protegidas;

28. Realizar a recuperação das áreas eventualmente degradas causadas pelos Serviços das Obras de Implantação e Pavimentação da Rodovia;

29. Recuperar todo o passivo ambiental decorrente as implantação dos Serviços das Obras de Implantação e Pavimentação da Rodovia;

30. Comunicar previamente aos proprietários vizinhos sobre eventuais interferências / interdições nos acessos e em cercas existentes às margens da Rodovia, para evitar o cerceamento do-direito de ir e vir dos moradores vizinhos e a debandada de rebanhos das propriedades;

31. A execução das obras não poderá causar danos ao meio ambiente e a terceiros e, caso ocorra, acidentalmente ou não, a AGETOP deverá se responsabilizar tanto pela recuperação das áreas danificadas / atingidas, como por qualquer outra responsabilidade originada por sua má execução.

32. A SEMARH, isenta-se das obrigações com os proprietários que tiverem suas propriedades interceptadas pelo projeto, cabendo à AGETOP as tratativas com relação a desapropriações e indenizações;

33. Isolar a área sob influência direta das obras na etapa de sua implantação, objetivando controlar o acesso de terceiros;

34. Recuperar as caixas de empréstimos, as áreas de bota-fora de solos inadequados, áreas dos canteiros de obras, locais de armazenagem de combustíveis, pátios, acessos auxiliares e desvios, eventualmente utilizados e/ou implantados por ocasião da realização dos serviços das Obras de Implantação e Pavimentação da Rodovia;

35. Fazer a revegetação dos cortes, encostas e aterros visando a estabilidade dos mesmos;

36. Cumprir todas as medidas mitigadoras propostas no EIA/RIMA e nas normas e procedimentos ambiental e Planos Básicos para empreendimentos rodoviários;

37. Manter os motores a combustão dos equipamentos utilizados na obra - motoniveladoras, carregadeiras, escavadeiras, retroescavadeiras, rolos pés de carneiro, rolos compactadores, caminhões caçambas, caminhão comboio de lubrificantes/combustível, caminhão espargidor vibro acabadoras, rolos de pneus dentre outros, bem regulados e com emissão de gases poluentes nos padrões ambientais aceitáveis;

38. Não implantar canteiros de obras próximo a corpo hídrico, com vistas a evitar danos ambientais em Áreas de Preservação Permanente – APP;

39. Não efetuar quaisquer tipos de intervenções nas Áreas de Preservação Permanente, vertentes, nascentes, e áreas próximas a drenagens naturais sem a prévia autorização da SEMARH;

40. Não causar nenhum tipo de comprometimento nos vales e veredas localizadas na área de influência do empreendimento;

41. Não comprometer os cursos hídricos interceptados pela estrada, quanto aos aspectos quanti e qualitativos dos mesmos, bem como manter preservadas as matas ciliares;

42. Aterramento de cabeceiras ou mesmo de drenagens intermitentes não poderão ser efetuados neste projeto, uma vez que o referido empreendimento está na zona de amortecimento de um Parque Nacional;

43. Por tratar-se de obra de engenharia civil, manter acompanhamento técnico qualificado na sua implantação.

44. O Projeto deverá contemplar o aproveitamento da estrada existente, inibindo-se retificações de trechos, o que poderia acarretar supressão de vegetação devido às características florísticas da região;

45. Esta Licença Ambiental, não autoriza a supressão da Flora, que deverá ser devidamente licenciada nesta SEMARH/SULIM/GFF;

46. Esta Licença Ambiental não dispensa o licenciamento ambiental específico do(s) Canteiro(s) de Obras;

47. Esta Licença Ambiental não dispensa o licenciamento ambiental específico da(s) área(s) de bota fora de solos inservíveis;

48. Esta Licença Ambiental não dispensa o licenciamento ambiental específico da(s) Jazida(s) de cascalho ou de solos;

49. As áreas de empréstimos não poderão se localizar na área de influência do empreendimento, devendo primeiramente ser mapeadas e localizadas, em escala de 1:1000, observando a distância que se localizam do Parque Nacional Chapada dos Veadeiros. Essas áreas deverão ter licenciamentos específicos;

50. Esta Licença Ambiental não dispensa o licenciamento ambiental específico de Usinas de Asfalto de todo e qualquer tipo;

51. Esta Licença Ambiental não dispensa o licenciamento ambiental específico de Usinas de Concreto;

52. Esta Licença Ambiental não dispensa a Portaria específica de Outorga de Água;

53. A implantação de áreas de extração de areia, usinas de concreto e canteiros de obras, não deverão ocorrer nas APPs nem nas Reservas Legais e nem em matas nativas;

54. Apresentar após o início das obras, o Plano de Desmobilização da Implantação e Pavimentação da Rodovia;

55. A faixa de domínio da Rodovia não deverá adentrar os limites do Parque Nacional Chapada dos Veadeiros. O Plano Básico Ambiental deverá incluir mapa em escala mínima de 1:5000, com as delimitações da Unidade de Conservação georreferenciadas, a rodovia e a sua faixa de domínio, bem como os limites de zona de amortecimento, segundo resolução CONAMA 013/1990;

56. Manter sempre, uma cópia desta LICENÇA DE INSTALAÇÃO das Obras da Rodovia, no(s) canteiro(s) de obras, frentes de serviços, escritórios da firma contratada e departamento de meio ambiente da requerente, para efeitos de fiscalizações e vistorias da SEMARH;

57. Apresentar Relatório de Acompanhamento/Monitoramento Ambiental na finalização da obra, elaborado por no mínimo dois técnicos habilitados, acompanhado de relatório fotográfico;

58. Após o início dos serviços das obras, apresentar a LEF - Licença de Exploração Florestal;

59. A rodovia não poderá ter "pontos de parada" para turistas, com exceção do já existente "Jardim Maytréa", que deverá receber apenas tratamento de contenção de processos erosivos;

60. Para pavimentação da rodovia deverá ser utilizada toda a tecnologia conhecida para que a coloração do asfalto seja a mais clara possível;

61. Nos pontos selecionados pelo estudo de impacto ambiental como corredores de fauna, deverão existir redutores de velocidade e velocidade máxima de 40 km/h, além dos sonorizadores de ambos os lados, os quais deverão ser detalhados no PBA. A sinalização da existência de Parque Nacional e informação de trânsito de animais deverá ser mais requente;

62. Na área de influência direta do Parque a velocidade máxima permitida na rodovia não deverá ser superior a 620km/h, devido ao tempo necessário de frenagem, caso haja animais na pista;

63. Em drenagens que se caracterizam com corredores de fauna deverão ser programados bueiros com o mínimo de 2,00 metros de diâmetro;

64. Sugere-se ainda, que as rodovias que dão acesso a BR-153, ou mesmo para o norte de Goiás (como a de Padre Bernardo, p. ex.) seja integralmente asfaltadas ou pavimentadas, em curto prazo, para que o fluxo de veículos possa ser melhor distribuído, não sobrecarregando a GO-239 que margeia o Parque Nacional

Chapada dos Veadeiros;

65. Encaminhar a SEMARH, o relatório de controle ambiental – RCA das medidas implantadas;

66. Por tratar-se de uma Estrada Parque, a velocidade máxima na rodovia deverá ser de 60km/h;

67. Redução da velocidade, nos pontos onde se constatou a travessia de animais para 40km/h;

68. Fica creditada aos responsáveis técnicos que elaboraram o EIA/RIMA as viabilidades técnicas ambientais pelas Obras de Implantação, Restauração, Readequação, Obras de Arte Corrente, Obras de Artes Especiais, Pavimentação, Sinalização Vertical e Horizontal da Rodovia GO-239, no Trecho de Alto Paraíso até o Distrito de São Jorge e a cidade de Colinas do Sul, com extensão de 70 quilômetros e outras dela decorrentes;

69. Apresentar Plano de Recuperação de Áreas Degradadas – PRAD;

70. Esta licença está sendo concedida com base nas informações e documentos anexados ao processo, entendendo-se os mesmos como verídicos, sabendo-se que a inveracidade nos mesmos culminará no cancelamento da presente licença;

71. Esta Secretaria reserva o direito de fazer novas exigências caso considere necessário.

Exigências Técnicas de Compensação Ambiental SNUC/SEUC

1. Referência Parecer Nr. 30317/2013, elaborado por Marco Antônio Asevedo Brito

2. Para cumprimento da compensação ambiental prevista na Lei Federal 9.985/2000 e Lei Estadual 14.247/2002, foi celebrado o Termo de Compromisso de Compensação Ambiental SNUC/SEUC número: 02/13

Exigências Técnicas de Compensação Ambiental de Fauna

1. Referência Parecer Nr. 30316/2013, elaborado por Marco Antônio Asevedo Brito

2. Para cumprimento da compensação ambiental prevista na Lei Estadual 14.241/2002, foi celebrado o Termo de Compromisso de Compensação Ambiental Fauna número: OFICIO AGETOP 304/2013-DEP de 26.08.2013

Nota

1. A AGETOP, deverá atender às recomendações constantes do Parecer Técnico Parque Nacional Chapada dos Veadeiros da Coordenação de Assistência Técnica e Educação Ambiental 06/06 de 26.12.2006, folhas 286-299 do processo;

2. As gramíneas a serem utilizadas nas proteções de taludes de cortes e de aterro, deverão ser preferencialmente as do tipo Paspalum notatum, vulgarmente conhecida como Grama Batatais, conforme recomendação do Ofício 087/04-PNCV de 03.11.2004;

3. Foram realizados até o momento a pavimentação de 23 quilômetros, trecho esse que necessita de restaurações e adequações.

4. Analista Ambiental Engenheiro Civil Marco Antonio Asevedo Brito.

Validade da Licença: 27/08/2015

Goiânia, 27/08/2013.

José Augusto dos Reis Cruz
Gerente
GERÊNCIA DE USO DO SOLO

Elaboração:

Pryscila Teixeira Margon
Gestor Público
SUPERINTENDÊNCIA DE LICENÇA E MONITORAMENTO

Buy your books fast and straightforward online - at one of world's fastest growing online book stores! Environmentally sound due to Print-on-Demand technologies.

Buy your books online at
www.morebooks.shop

Kaufen Sie Ihre Bücher schnell und unkompliziert online – auf einer der am schnellsten wachsenden Buchhandelsplattformen weltweit! Dank Print-On-Demand umwelt- und ressourcenschonend produzi ert.

Bücher schneller online kaufen
www.morebooks.shop

Printed by Books on Demand GmbH, Norderstedt / Germany